Sustainable River Basin Management: Strategies for Climate Resilience, Water Security, and Ecosystem Conservation

Introduction to Sustainable River Basin Management

Chapter 1: The Hydrological Cycle and River Basin Functionality

Chapter 2: Integrated River Basin Management

Chapter 3: Water Governance in River Basins

Chapter 4: Ecosystem-Based Approaches to River Basin Management

Chapter 5: Climate Change and River Basin Resilience

Chapter 6: Sustainable Water Use Across Sectors

Chapter 7: Pollution Management in River Basins

Chapter 8: Transboundary River Basin Management

Chapter 9: Stakeholder Engagement in River Basin Management

Chapter 10: Financing Sustainable River Basin Management

Conclusion: The Future of Sustainable River Basin Management

Introduction to Sustainable River Basin Management

Sustainable river basin management involves a coordinated, holistic approach to managing water resources across a river basin to ensure that natural ecosystems, water quality, and human needs are balanced. This integrated method takes into account the hydrological, ecological, and socio-economic dynamics within the river basin and emphasizes the importance of cooperation among stakeholders, such as governments, local communities, industries, and agricultural sectors. Given that river basins are complex and interconnected systems, sustainable management approaches focus on long-term resilience and equitable resource distribution to address issues like water scarcity, pollution, and ecosystem degradation.

River basins encompass the entirety of the area where water flows toward a main water body, like a river or lake, and include elements such as tributaries, groundwater, floodplains, and estuaries. This geographic perspective allows for the consideration of all sources and uses of water within the basin, encouraging policies and practices that go beyond mere water management to include land use, pollution control, and biodiversity conservation. Sustainable river basin management is fundamentally about achieving a balance between human and environmental needs, maintaining water quality, and enhancing ecosystem health to ensure that rivers and watersheds can continue to support diverse uses now and in the future.

Definition and Scope of Sustainable River Basin Management

River basins, or watersheds, are natural boundaries for managing water resources, and sustainable management within these areas requires understanding their environmental, social, and economic dimensions. Sustainable river basin management is not limited to water quantity; it also emphasizes water quality, biodiversity, soil health, and the ecosystems within the basin. For instance, land use activities, such as agriculture, urbanization, and industry, affect water availability and quality, impacting downstream users and ecosystems. Therefore, sustainable river basin management includes practices that maintain the river basin's hydrological and ecological balance while supporting human needs and economic activities.

The scope of sustainable river basin management spans technical aspects, such as water flow regulation and pollution control, and involves legal frameworks, stakeholder engagement, and capacity building to address competing demands and complex governance issues. A key feature of sustainable river basin management is its integrated perspective: it views water as part of a larger system that includes land, flora, fauna, and human communities. This approach addresses challenges like land degradation, deforestation, and the effects of industrial activities on river ecosystems.

Importance of Integrated Watershed Management for Long-Term Sustainability

Integrated Watershed Management (IWM) is fundamental to achieving sustainable river basin management. By managing land and water resources together, IWM enables a coordinated approach to address the ecological, social, and economic factors influencing the health and resilience of a watershed. It combines land-use planning, pollution control, conservation, and water resource management, treating the river basin as a single, interconnected system.

IWM promotes sustainable land and water use practices that enhance ecosystem services, such as natural water filtration, flood mitigation, soil health, and habitat preservation. These services are essential for maintaining water quality and quantity for human and ecological needs. Furthermore, IWM supports resilience-building by allowing managers to adapt to changes caused by climate fluctuations, population growth, and economic development. The approach fosters collaboration among various stakeholders, creating a platform for cooperation to align individual and collective goals with the long-term health of the basin.

The resilience and adaptability provided by IWM are crucial in a changing world, where climate variability and anthropogenic pressures threaten traditional water availability and quality patterns. By adopting IWM principles, river basin managers can implement strategies that reduce vulnerability to floods, droughts, and pollution, contributing to the sustainability of ecosystems and communities reliant on river basins.

Relevance of Sustainable River Basin Management in the Context of Climate Change and Water Security

Climate change poses significant challenges for river basin management by altering precipitation patterns, increasing the frequency of extreme weather events, and impacting water flows. These changes threaten both water quantity and quality, affecting agriculture, drinking water, industrial processes, and overall regional stability. River basin managers face the challenge of adapting to these shifts to safeguard water security and ecological health within basins. This requires flexible management strategies that incorporate climate change adaptation, risk management, and resilience.

River basins play a central role in water security, providing water resources for drinking, irrigation, energy, and industry. As climate change intensifies, traditional sources of water may become less reliable, necessitating sustainable river basin management to balance competing needs. Effective management strategies within basins can improve water quality, conserve critical ecosystems, and ensure equitable resource distribution, promoting peace and reducing competition over scarce resources.

Sustainable river basin management, therefore, is essential in building resilience to climate-induced stresses and supporting socio-economic stability. Through adaptive management practices, pollution control, and equitable water allocation, sustainable river basin management enables societies to secure essential water resources for current and future generations.

Challenges in River Basin Management

Managing a river basin sustainably presents various challenges, as these environments are highly interconnected and subject to both natural variability and human pressures. Key challenges include water scarcity, pollution, governance complexity, and climate change impacts. Understanding these challenges is essential to identifying sustainable solutions that address root causes rather than merely treating symptoms.

Water Scarcity, Pollution, and Over-Extraction

Water scarcity is a pressing issue in many river basins, exacerbated by population growth, industrial demands, and agricultural needs. Over-extraction of water resources often leads to reduced river flows, depleted aquifers, and loss of habitat for aquatic species. Pollution from agricultural runoff, industrial waste, and urban wastewater compounds the issue, deteriorating water quality and endangering public health, ecosystems, and food security.

The cumulative effects of water scarcity, pollution, and over-extraction place river basins under immense stress, threatening the sustainability of water resources. Sustainable management practices, including efficient water use, pollution control, and restoration efforts, are vital in reversing these trends and maintaining ecological balance within river basins.

Governance Issues and Competing Stakeholder Interests

River basin management involves numerous stakeholders, including governments, local communities, industries, and environmental organizations. These stakeholders often have conflicting interests, with some prioritizing economic development over environmental preservation, while others advocate for strict conservation measures.

Effective governance is critical to aligning these interests, yet achieving coordination and agreement is challenging due to diverse priorities and, in some cases, lack of institutional capacity.

Governance issues are particularly pronounced in transboundary river basins, where political boundaries divide resources that cross multiple jurisdictions. Disputes over water allocation, pollution control, and conservation measures can hinder cooperative efforts, leading to unsustainable practices. Addressing these governance challenges requires transparent decision-making, inclusive stakeholder engagement, and conflict resolution mechanisms.

Impact of Climate Change on River Basins (Droughts, Floods, Changing Precipitation Patterns)

Climate change exacerbates existing pressures on river basins, altering precipitation patterns, increasing temperatures, and causing more frequent extreme weather events like droughts and floods. These changes disrupt natural water cycles, reduce water availability, and increase the likelihood of extreme events that can damage infrastructure, erode land, and reduce agricultural productivity.

Sustainable river basin management aims to build resilience to these climate impacts by incorporating adaptive strategies such as floodplain restoration, reforestation, and sustainable land use practices. These measures help buffer against extreme events, maintain water quality, and support biodiversity, providing critical stability in an uncertain future.

Goals of the Book

This book provides a comprehensive overview of sustainable river basin management, offering insights into best practices and innovative strategies to manage water resources effectively. By exploring these strategies, readers will gain an understanding of the complexities and importance of river basin management in a changing world.

To Provide a Comprehensive Overview of Sustainable Practices in River Basin Management

The book aims to cover a range of sustainable practices that address the unique challenges of managing river basins. This includes discussions on integrated watershed management, ecosystem-based approaches, pollution control, and strategies for stakeholder engagement. Each chapter will delve into specific issues, providing a foundation of knowledge for practitioners, policymakers, and researchers interested in sustainable water management.

To Discuss Strategies for Improving Water Governance, Ecosystem Health, and Water Resource Allocation

A key focus of the book is to explore the governance structures, policies, and institutional arrangements that support sustainable river basin management. This includes examining successful case studies, as well as identifying gaps and barriers to effective management. By highlighting strategies that improve water governance and ecosystem health, the book offers actionable insights for enhancing river basin resilience.

To Present Solutions for Sustainable Water Use Across Sectors within a River Basin

Water demand spans multiple sectors, including agriculture, industry, and domestic use. This book seeks to present strategies that encourage sustainable water use across these sectors while protecting water resources. It highlights cross-sectoral approaches that balance economic needs with environmental protection, promoting a future where all users within a river basin can access and benefit from clean, safe, and adequate water resources.

In summary, this book underscores the critical role of sustainable river basin management in addressing global water challenges. Through an integrated and adaptive approach, it provides the tools and knowledge necessary to protect river basins, supporting both human well-being and ecosystem health in an era of unprecedented environmental change.

Chapter 1: The Hydrological Cycle and River Basin Functionality

River basins, also known as watersheds, are geographic areas that collect precipitation and direct it toward a specific water body, typically a river, lake, or ocean. These areas serve as natural boundaries for water resource management, offering a holistic view of water movement, storage, and utilization. Sustainable river basin management relies heavily on an in-depth understanding of river basin structures and the processes within the hydrological cycle. River basins encompass diverse ecosystems, ranging from headwaters to floodplains, each serving unique roles in water flow regulation, ecosystem support, and human utilization.

In this chapter, we explore the key elements of river basin functionality and the critical role of the hydrological cycle. The chapter will outline the structure of river basins, the processes within the hydrological cycle, and the valuable ecosystem services that river basins provide. Finally, we will examine the impacts of human activities, such as deforestation, urbanization, and industrialization, on river basin health. This foundational knowledge is essential for understanding how sustainable river basin management can protect water resources, support biodiversity, and enhance resilience against environmental challenges.

Introduction to River Basins

River basins represent the most effective natural boundary for water resource management, encompassing an area where all precipitation, surface water, and groundwater flow toward a common outlet. The concept of a river basin allows for an integrated approach to managing water resources, ecosystems, and human activities within defined geographical boundaries. By examining the components of a river basin and their functions, we gain insights into the complexity and interconnectedness of water systems.

Definition and Structure of a River Basin

A river basin is a geographical area drained by a river and its tributaries, defined by the topography of the landscape. River basins vary widely in size, from small watersheds that feed into local streams to expansive basins like the Amazon, Mississippi, or Nile, which span multiple countries and regions. Each basin has natural boundaries, often determined by high elevations such as mountains or ridges, where precipitation collects and begins its journey through the basin. River basins are divided into sub-basins or catchments, which capture and direct water toward the main river channel.

A comprehensive understanding of a river basin's structure is essential for effective water management. River basins provide natural infrastructure for directing water flow, maintaining water quality, and supporting ecosystems. This structure includes several key components, each of which plays a vital role in water flow regulation, nutrient cycling, and ecosystem health.

1. Headwaters: The headwaters, or source, are the highest elevation points in a river basin where precipitation accumulates and initiates water flow. This area often includes small streams, springs, or glaciers that supply the river with a continuous source of water. The headwaters are crucial for sustaining flow levels in downstream areas, especially during dry seasons.

2. Tributaries: Tributaries are smaller streams or rivers that feed into the main river channel, collecting surface runoff, groundwater discharge, and nutrients from surrounding landscapes. Tributaries play a critical role in transporting water and organic matter, contributing to the nutrient cycles that support aquatic ecosystems. In many river basins, the network of tributaries spans a vast area, creating a complex web of interconnected water sources.

3. Main River: The main river is the primary watercourse within a river basin, carrying water, sediments, and nutrients toward the basin's outlet. This central artery connects different parts of the

basin, facilitating the movement of species, nutrients, and sediments across diverse ecosystems. The main river serves as a vital water source for human communities, agriculture, and industry, highlighting the need for sustainable management to balance these uses with ecosystem health.

4. Floodplains: Floodplains are low-lying areas adjacent to rivers that store excess water during high-flow events, such as seasonal floods. These areas serve as natural buffers, reducing flood risk in downstream communities and supporting biodiversity with nutrient-rich soils. Floodplains provide essential ecosystem services, including soil formation, habitat for wildlife, and water filtration.

5. Estuaries: Estuaries are transition zones where freshwater from rivers meets saltwater from the ocean, creating unique and highly productive ecosystems. Estuaries support diverse plant and animal species, serving as nurseries for fish and other marine life. Additionally, estuaries act as natural filters, trapping sediments, nutrients, and pollutants before they reach the ocean. Protecting estuarine ecosystems is vital for preserving water quality, biodiversity, and coastal resilience.

Each component within a river basin plays a role in regulating water flow, maintaining ecosystem health, and supporting human use. Understanding the structure of river basins is foundational to sustainable management practices, as it highlights the interconnectedness of natural processes and the importance of preserving each component's functionality.

The Role of the Hydrological Cycle in River Basins

The hydrological cycle, or water cycle, is the driving force behind water movement within river basins. This continuous process involves precipitation, runoff, infiltration, groundwater recharge, evaporation, and transpiration. The hydrological cycle not only sustains water availability but also influences nutrient cycles, sediment transport, and habitat conditions within a river basin. Each stage of the hydrological cycle plays a role in distributing water resources and maintaining ecological balance.

1. Precipitation: Precipitation is the primary input of water into a river basin, occurring as rain, snow, sleet, or hail. The amount, timing, and distribution of precipitation directly impact river flows, groundwater recharge, and water availability. Seasonal precipitation patterns, such as monsoons or winter snowmelt, create distinct hydrological dynamics within river basins. For example, snowmelt in mountainous areas can contribute to river flow during warmer months, sustaining water levels when precipitation is limited.

2. Runoff: After precipitation, water that does not infiltrate into the soil flows over the land surface as runoff, moving toward rivers, lakes, and other water bodies. Runoff transports sediments, nutrients, and organic matter into river basins, supporting nutrient cycles and aquatic ecosystems. However, excessive runoff can lead to soil erosion, sedimentation, and water quality degradation, especially when influenced by human activities like agriculture and urbanization.

3. Infiltration and Groundwater Recharge: Infiltration occurs when water seeps into the soil, replenishing underground water reserves, or aquifers. Groundwater recharge is essential for maintaining base flow levels in rivers, particularly during dry seasons. The rate of infiltration and groundwater recharge depends on factors such as soil type, vegetation cover, and land use. Healthy forests and grasslands, for instance, promote infiltration, reducing runoff and supporting groundwater recharge.

4. Surface Water and Groundwater Interactions: Surface water bodies, like rivers and lakes, are often connected to groundwater systems through natural seepage and discharge processes. Groundwater can maintain river flow levels during dry periods, providing a steady water supply for ecosystems and human communities. However, excessive groundwater extraction for

agriculture or industry can lower water tables, disrupt flow patterns, and reduce base flows, affecting the availability of surface water.

The hydrological cycle is the underlying mechanism that regulates water distribution within a river basin. By maintaining the balance between precipitation, runoff, infiltration, and groundwater recharge, the hydrological cycle supports water availability, quality, and ecosystem health. Understanding these processes is essential for sustainable river basin management, as it highlights the need to preserve both surface and groundwater resources.

Ecosystem Services Provided by River Basins

River basins provide a wide array of ecosystem services—benefits that sustain human societies and the natural environment. These services include biodiversity support, water purification, soil formation, flood regulation, and socioeconomic resources. Recognizing and preserving these services is vital for promoting resilience and sustainability within river basins.

1. Biodiversity Support: River basins host diverse ecosystems, from mountain streams and wetlands to floodplains and estuaries. These habitats support a wide variety of plant and animal species, contributing to global biodiversity. Biodiversity within river basins enhances ecosystem resilience, allowing species to adapt to environmental changes and disturbances. Conserving biodiversity in river basins is crucial for maintaining ecological stability and the services these ecosystems provide.

2. Water Purification: Natural processes within river basins, such as filtration through vegetation and wetlands, help purify water. Wetlands act as natural filters, trapping pollutants, sediments, and nutrients before they reach rivers. Riparian vegetation along riverbanks absorbs excess nutrients from agricultural runoff, improving water quality. These natural purification processes reduce the need for costly water treatment infrastructure and provide clean water for drinking, irrigation, and industry.

3. Soil Formation and Nutrient Cycling: River basins play a significant role in soil formation and nutrient cycling, which are essential for agricultural productivity and ecosystem health. Floodplains receive periodic sediment deposits during floods, enriching soils with nutrients. This natural fertilization supports crop growth and enhances soil fertility. The nutrient cycles within river basins contribute to sustainable agriculture, food security, and biodiversity.

4. Flood Regulation: Floodplains and wetlands within river basins store excess water during high-flow events, reducing downstream flood risk. These areas act as natural buffers, managing floodwaters and protecting communities and infrastructure. Flood regulation is particularly valuable in regions prone to seasonal flooding, where natural floodplains can mitigate the impacts of extreme weather events. Preserving floodplain ecosystems is crucial for maintaining the basin's natural capacity to absorb and store floodwaters.

5. Socioeconomic Value: River basins support economic activities such as agriculture, fisheries, hydropower, and tourism. Fertile soils and reliable water supplies enable productive agriculture, providing food and income for local populations. Fisheries in rivers and estuaries offer a source of protein and livelihoods, while hydropower generates renewable energy for urban and rural areas alike. Additionally, scenic river landscapes attract tourists, creating revenue and employment opportunities for nearby communities.

The ecosystem services provided by river basins are essential for human well-being and environmental sustainability. Sustainable river basin management aims to protect and enhance these services, recognizing their role in supporting biodiversity, economic stability, and resilience to climate change.

Impacts of Human Activities on River Basin Functionality

Human activities within river basins significantly impact water quality, flow patterns, and ecosystem health. Unsustainable practices, such as deforestation, urbanization, agriculture, and industrialization, can degrade river basin functionality, leading to water scarcity, pollution, and biodiversity loss. Understanding these impacts is essential for implementing management strategies that mitigate harm and promote sustainable use of river basin resources.

1. Deforestation: Deforestation disrupts the hydrological cycle within river basins by removing trees and vegetation that intercept rainfall, reduce runoff, and promote groundwater recharge. When forests are cleared, more rainfall becomes surface runoff, increasing the risk of soil erosion and sedimentation in rivers. Deforestation also reduces the capacity of river basins to regulate water flows, making downstream areas more vulnerable to floods and droughts.

2. Urbanization: As populations grow, urban expansion places pressure on river basins by increasing demand for water, generating waste, and altering land use. Urban areas often contain impermeable surfaces, like roads and buildings, which prevent water infiltration and accelerate runoff. This can lead to more frequent and severe flooding and reduced groundwater recharge. Additionally, urban wastewater discharges can introduce pollutants into rivers, degrading water quality and harming aquatic life.

3. Agriculture: Agriculture is a major driver of environmental change in river basins, as irrigation, fertilization, and pesticide use impact water availability and quality. Irrigation diverts large quantities of water from rivers, reducing flow levels and affecting downstream ecosystems. Fertilizers and pesticides contribute to nutrient pollution and chemical contamination, which impair water quality and harm aquatic biodiversity. Sustainable agricultural practices, such as conservation tillage and buffer zones, can help mitigate these impacts.

4. Industrialization: Industrial activities within river basins often require substantial water resources and produce waste byproducts

that can pollute water bodies. Industries such as mining, manufacturing, and energy generation may discharge chemicals, heavy metals, and thermal pollution into rivers, affecting water quality and posing risks to public health and wildlife. Implementing pollution controls, adopting cleaner production methods, and enforcing regulations are critical steps for reducing the environmental footprint of industries in river basins.

5. Alteration of Natural Flow Patterns: Human activities frequently alter the natural flow regimes of rivers, impacting ecosystems and reducing water availability for downstream users. Dams, for example, disrupt natural flow patterns by storing and releasing water according to human demands rather than seasonal variations. Channelization—the process of straightening and deepening rivers for navigation or flood control—further modifies flow patterns and reduces habitat complexity.

The alteration of natural flow patterns and degradation of ecosystem services are significant consequences of human activities within river basins. Sustainable river basin management seeks to balance human needs with the preservation of natural functions, implementing practices that minimize harm and support the resilience of both ecosystems and communities.

Conclusion

Understanding the hydrological cycle and the functionality of river basins is fundamental to sustainable water management. River basins are complex systems that provide essential ecosystem services, from biodiversity support and water purification to flood regulation and economic resources. However, human activities can severely impact these functions, highlighting the importance of adopting sustainable practices that protect and restore river basin health. By recognizing the interconnectedness of hydrological processes, ecosystems, and human influences, sustainable river basin management can promote resilience, adaptability, and equitable resource use within these vital landscapes.

Chapter 2: Integrated River Basin Management

Integrated River Basin Management (IRBM) provides a holistic framework that addresses the complex interplay between water, land, and ecosystems within a river basin. Traditional water management approaches often focus narrowly on isolated aspects, potentially overlooking the interconnected dynamics of river systems, ecological functions, and human activities. IRBM, by contrast, takes a systems-based approach, treating the entire river basin as a single, interconnected unit. This approach allows for a more sustainable management strategy, emphasizing the ecological, social, and economic dimensions of water resources. As global water resources come under increasing pressure from climate change, pollution, population growth, and expanding human activities, IRBM has emerged as a vital approach to ensure that water resources are managed in ways that support both people and ecosystems over the long term.

In this chapter, we will examine the foundational principles guiding IRBM, focusing on the necessity of a multi-stakeholder approach, the integration of diverse water uses alongside ecosystem needs, and the need for coordination across administrative, sectoral, and geographical boundaries. Next, we delve into the core components required to implement IRBM effectively, from resource assessment and monitoring to policy frameworks and stakeholder engagement. Lastly, we present real-world examples of IRBM in action, including successes and challenges faced in managing the Danube and Rhine river basins. By understanding these examples, we gain insights into the potential of IRBM to foster resilience, sustainability, and cooperation within shared water systems.

Principles of Integrated River Basin Management

Integrated River Basin Management is based on foundational principles that set it apart from more traditional, fragmented water management approaches. These principles serve as guidelines to ensure that all aspects of the river basin, including ecological integrity and human needs, are considered and integrated into management decisions.

Multi-Stakeholder Approach

At the heart of IRBM is the principle that all stakeholders within a river basin must have a voice in its management. River basins are often home to a wide range of stakeholders, each with unique needs, perspectives, and levels of influence. These stakeholders include government agencies, local communities, agricultural users, industrial sectors, environmental groups, and international organizations in cases where the basin spans multiple countries. The multi-stakeholder approach ensures that diverse interests are acknowledged and balanced, reducing the likelihood of conflict and promoting cooperation. Engaging a wide array of stakeholders fosters transparency and inclusivity, which are essential for building trust, sharing responsibilities, and fostering a sense of shared ownership over the basin's resources. This approach also allows for more innovative solutions, as stakeholders can pool resources, knowledge, and expertise to address shared challenges.

Consideration of All Water Uses and Ecosystem Needs

Unlike conventional water management methods that prioritize human water demands, IRBM emphasizes the need to balance these demands with the ecological needs of the river basin. Water in a river basin serves multiple purposes, supporting human consumption, agriculture, industry, energy generation, and recreation. Simultaneously, it sustains diverse ecosystems, from wetlands and riparian zones to floodplains and estuaries, all of which provide essential services, such as water purification, habitat for wildlife, nutrient cycling, and flood mitigation. IRBM integrates these diverse requirements, aiming to allocate water in ways that benefit both people and nature. By ensuring that ecosystems have sufficient water, IRBM helps maintain biodiversity, enhance resilience, and provide a natural buffer against environmental fluctuations, such as droughts and floods. This balance is critical for long-term sustainability, as healthy ecosystems are more capable of adapting to changing conditions and continuing to provide the services on which human populations depend.

Coordination Across Administrative, Sectoral, and Geographical Boundaries

River basins frequently span multiple administrative jurisdictions, including local, regional, national, and even international boundaries. Managing a river basin that crosses these various jurisdictions presents challenges, as different areas may have distinct regulations, priorities, and resources. IRBM addresses these challenges by promoting coordination across boundaries, encouraging collaboration between different sectors, administrations, and regions that share the basin. Coordination ensures that policies and actions are aligned, minimizing the risk that measures taken in one area will have adverse impacts on others. In transboundary river basins, this principle is especially important, as upstream activities can significantly affect downstream water quality, quantity, and ecosystem health. By fostering partnerships and aligning policies across borders, IRBM reduces conflicts and supports a more cohesive, cooperative approach to shared water resources.

The principles of IRBM provide a comprehensive foundation for sustainable water resource management within a river basin. By promoting stakeholder engagement, balancing human and ecological needs, and fostering cross-boundary coordination, these principles ensure that IRBM is a flexible and inclusive approach capable of addressing the complexities of modern water management.

Key Components of IRBM

The implementation of IRBM relies on several key components that translate its guiding principles into actionable strategies. These components allow water managers to assess resources accurately, make informed decisions, and engage stakeholders effectively, thus ensuring that management practices are both inclusive and sustainable.

Water Resource Assessment and Monitoring

Accurate data on water resources is essential for effective IRBM. Water resource assessment and monitoring involve collecting data on water quantity, quality, seasonal variations, groundwater levels, and sources of pollution. This data forms the basis for informed decision-making, allowing stakeholders to understand the current state of the river basin and to track changes over time. Regular monitoring supports adaptive management by providing insights into the effectiveness of policies and interventions, enabling managers to adjust strategies as needed. Advances in technology, such as geographic information systems (GIS), remote sensing, and hydrological modeling, have greatly improved the ability to monitor and assess water resources at fine scales. Continuous monitoring allows water managers to detect emerging threats early, respond to changes in water availability, and allocate resources in a way that aligns with sustainability goals. These data-driven insights are particularly important in the context of climate change, where shifts in precipitation patterns and extreme weather events can have significant impacts on water resources.

Policy Frameworks and Legal Instruments

IRBM requires a strong policy and legal framework that defines the rules, rights, and responsibilities associated with water use within the river basin. Policy frameworks establish guidelines for sustainable water allocation, pollution control, land use, and environmental protection, creating a regulatory basis for IRBM. Effective policy frameworks align with the principles of IRBM, promoting coordination across sectors, engagement of stakeholders, and the protection of ecosystem services. Legal instruments, such as water rights, pollution permits, and environmental regulations, provide enforceable mechanisms that ensure compliance with sustainable practices. These legal frameworks help prevent conflicts, provide a basis for dispute resolution, and hold stakeholders accountable for their actions. A robust policy and legal foundation is critical for the long-term success of IRBM, as it provides clarity and consistency in managing water resources within the basin.

Public Participation and Stakeholder Engagement

Engaging stakeholders is a central component of IRBM, as it fosters inclusivity, transparency, and shared responsibility. Public participation allows local communities, industries, environmental groups, and other stakeholders to voice their concerns, share knowledge, and contribute to decision-making processes. Stakeholder engagement can take various forms, including public consultations, joint planning sessions, and community-led water management initiatives. By involving stakeholders in discussions about water management, IRBM fosters a sense of ownership and accountability among participants, making them more likely to support and comply with policies. Public participation also enhances IRBM's adaptability, as local knowledge provides valuable insights into environmental changes, cultural factors, and social dynamics that may not be apparent from scientific data alone. This engagement ensures that management decisions are culturally sensitive, economically viable, and ecologically sound, reflecting the diverse perspectives of those who depend on the river basin.

Integrated Planning and Adaptive Management

Recognizing that river basins are complex and constantly evolving, IRBM emphasizes integrated planning and adaptive management. Integrated planning brings together water management, land-use planning, climate adaptation, and disaster risk reduction, creating a comprehensive approach to managing the basin's resources. Adaptive management allows water managers to modify strategies based on new information, changing conditions, or the observed outcomes of previous interventions. This flexibility is essential for addressing uncertainties, such as population growth, economic development, and climate variability, which can significantly affect water resources. By allowing for adjustments and real-time responses, adaptive management enables IRBM to remain relevant and effective in the face of changing circumstances, enhancing the resilience of both human communities and ecosystems within the basin.

These components—resource assessment, policy frameworks, stakeholder engagement, and adaptive management—provide the foundation for successful IRBM implementation. Together, they enable water managers to create a sustainable, inclusive, and resilient approach to managing the resources of a river basin.

Success Stories in IRBM

Several river basins around the world serve as examples of how IRBM principles and components can be applied to address complex water management challenges. These success stories illustrate the potential of IRBM to foster sustainability, resilience, and cooperation in shared water systems.

Danube River Basin

The Danube River Basin, which spans 19 countries, is one of the world's most complex transboundary river basins. The Danube River Protection Convention and the International Commission for the Protection of the Danube River (ICPDR) have established a cooperative framework for managing the basin, promoting joint water quality monitoring, pollution control, and habitat restoration. Through this framework, the ICPDR engages a wide range of stakeholders, including national governments, regional authorities, local communities, and non-governmental organizations. By coordinating actions and fostering collaboration, the ICPDR has achieved significant improvements in water quality, reductions in pollutant levels, and restoration of degraded ecosystems. The Danube IRBM model demonstrates how a commitment to multi-stakeholder engagement, cross-border cooperation, and environmental stewardship can yield positive outcomes, even in a complex and politically sensitive transboundary setting.

Rhine River Basin

The Rhine River Basin, which spans six European countries, provides another example of successful IRBM implementation. Managed by the International Commission for the Protection of the Rhine (ICPR), IRBM efforts in the Rhine Basin have focused on reducing pollution, managing flood risks, and restoring biodiversity. Over the past several decades, these initiatives have led to substantial improvements in water quality, with pollutant levels declining and aquatic species rebounding. The ICPR's success in coordinating policies across borders, engaging diverse stakeholders, and promoting sustainable practices underscores the potential of IRBM to foster resilience and sustainability within a shared river basin. The Rhine model highlights the importance of long-term commitment, adaptive management, and international cooperation in achieving lasting environmental and social benefits.

Lessons Learned from Failures and Challenges

Despite its successes, IRBM has also faced challenges that offer valuable lessons. In some cases, limited funding, political will, or stakeholder engagement has hindered IRBM initiatives. For example, insufficient integration of land-use planning with water management can lead to habitat loss, soil erosion, and degraded water quality. Additionally, competing interests among stakeholders can create obstacles to cooperation, particularly in regions where water scarcity intensifies competition for resources. These challenges highlight the need for strong institutional support, flexible funding mechanisms, and effective conflict resolution strategies to ensure the success and sustainability of IRBM initiatives. By learning from both successes and challenges, IRBM practitioners can refine their approaches and develop more resilient and adaptive management practices.

Through these examples, we gain a deeper understanding of IRBM's strengths and limitations. The successes of the Danube and Rhine river basins demonstrate IRBM's potential to improve water quality, enhance ecosystem resilience, and foster cross-boundary cooperation, while lessons from challenges emphasize the importance of adaptability, stakeholder engagement, and robust governance.

Conclusion

IRBM provides a comprehensive framework for sustainable water management within a river basin. By adopting a multi-stakeholder approach, balancing diverse water uses, and promoting cross-sectoral coordination, IRBM addresses the complexities of managing shared water resources in a way that considers both human and ecological needs. The principles of IRBM—such as inclusivity, ecological consideration, and cross-boundary collaboration—lay the foundation for addressing interconnected challenges within a river basin, supporting long-term sustainability and resilience.

The essential components of IRBM, including water resource assessment, policy frameworks, stakeholder engagement, and adaptive management, provide practical tools for implementing IRBM. These components enable water managers to make informed decisions, respond to changing conditions, and garner the support of diverse stakeholders. The success stories of the Danube and Rhine river basins illustrate IRBM's effectiveness in improving water quality, conserving biodiversity, and building resilience, while lessons from challenges underscore the importance of flexibility, conflict resolution, and institutional support.

In conclusion, IRBM represents a viable solution for promoting sustainable water management, ensuring that river basins continue to provide essential resources and ecosystem services for future generations. By building on IRBM principles and practices, water managers can create resilient, adaptive management strategies that protect the health and sustainability of river basins in the face of increasing environmental pressures.

Chapter 3: Water Governance in River Basins

Effective water governance within river basins is fundamental to ensuring the sustainable management of water resources. Governance defines the structures, policies, and processes that determine how water is managed, who has access to it, and how decisions are made regarding its allocation and protection. Without a strong governance framework, even well-planned water management strategies can falter, as fragmented responsibilities, inadequate policies, and lack of stakeholder coordination hinder effective decision-making and resource allocation.

This chapter explores the essential role of governance in river basin management. We begin by examining the importance of good water governance, addressing common challenges such as fragmented responsibilities and insufficient policy frameworks. Following this, we discuss water governance frameworks at different levels, from local to international, highlighting the need for multiscale cooperation in transboundary river basins. We then delve into the importance of institutional capacity building, emphasizing the need to strengthen institutions and equip local communities and stakeholders with the skills and knowledge required for sustainable water management. Finally, we present case studies of effective water governance initiatives, showcasing successful governance models that provide valuable lessons for river basin management worldwide.

The Importance of Good Water Governance

Good water governance is a cornerstone of sustainable river basin management, as it provides the structures, processes, and policies required to manage water resources effectively. Governance is not limited to governmental oversight; it includes the roles of all stakeholders—government agencies, local communities, industries, and non-governmental organizations—in decision-making processes. Governance frameworks determine how water is allocated, who has access, how conflicts are resolved, and how ecosystems are protected. Without good governance, even the best-planned management strategies may fail to achieve sustainability, as decisions are likely to be inconsistent, resources are misallocated, and conflicts may arise over access and usage.

Governance as a Critical Factor in Sustainable River Basin Management

Effective governance structures are essential for addressing the unique challenges of river basin management. River basins often span multiple administrative and jurisdictional boundaries, requiring coordinated decision-making across municipalities, regions, and, in some cases, national borders. In these complex systems, governance frameworks provide a structured approach for addressing competing demands on water resources, from agricultural irrigation and industrial use to household consumption and ecological preservation. Good governance ensures that decisions about water use and protection are transparent, equitable, and aligned with long-term sustainability goals. Moreover, governance frameworks provide mechanisms for conflict resolution, enabling stakeholders with differing interests to negotiate solutions that consider the needs of all parties involved.

Challenges in Water Governance: Fragmented Responsibilities and Inadequate Policies

Water governance within river basins is frequently challenged by fragmented responsibilities among different levels of government and various sectoral agencies. In many cases, responsibilities for water management are split between different authorities—environmental agencies, agricultural departments, municipal governments—leading to inconsistencies in decision-making and policy implementation. This fragmentation often results in conflicting regulations, inefficient resource use, and a lack of accountability. Additionally, water governance is frequently hampered by inadequate policy frameworks that fail to address the complexities of river basin management. In some regions, policies may lack provisions for stakeholder engagement, ecosystem protection, or transboundary cooperation, limiting the ability of governing bodies to manage resources effectively. Strengthening governance structures to overcome these challenges is essential for sustainable water management within river basins.

Good water governance provides the foundation for sustainable water resource management by ensuring that decisions are made transparently, equitably, and in a way that considers the long-term needs of both human and ecological systems. Addressing the challenges of fragmented responsibilities and inadequate policies is a critical step in building governance frameworks that support sustainable river basin management.

Water Governance Frameworks

Water governance frameworks vary significantly depending on the geographical scope, administrative structure, and specific needs of a river basin. Governance structures may operate at multiple levels—national, regional, and local—each playing a unique role in the management process. Additionally, transboundary river basins often require international cooperation to ensure that resources are managed equitably and sustainably across borders.

National, Regional, and Local Governance Structures

Effective water governance in river basins requires coordination across multiple levels of government. At the national level, governments are typically responsible for setting overarching policies, regulations, and standards for water use, quality, and conservation. These national frameworks provide a uniform approach to water management, ensuring that all regions within a country adhere to the same environmental and resource management standards. However, national policies must be flexible enough to account for the unique needs of specific river basins, as local ecological, social, and economic conditions can vary widely.

Regional governance structures, such as state or provincial agencies, often play an intermediary role, translating national policies into regionally appropriate management strategies. These agencies may be responsible for monitoring water quality, managing infrastructure, and overseeing land-use planning within the basin. Local governance structures, such as municipal governments and community organizations, are often closest to the issues and have a direct impact on the day-to-day management of water resources. They may oversee local water distribution, sanitation services, and community engagement in conservation efforts. Effective water governance frameworks integrate the roles of these different levels, promoting collaboration and communication among national, regional, and local stakeholders to ensure that water resources are managed comprehensively.

The Role of International Cooperation in Transboundary River Basins

Transboundary river basins present unique governance challenges, as water resources cross national borders and are subject to the jurisdiction of multiple countries. Effective governance in these basins requires international cooperation, as actions taken by one country can significantly impact water quality, availability, and ecosystem health downstream. International agreements, such as the United Nations Watercourses Convention and the Helsinki Rules on Transboundary Waters, provide frameworks for cooperation, emphasizing principles of equitable and reasonable use and the obligation to prevent harm to other states. These agreements encourage countries to negotiate water-sharing arrangements, establish joint monitoring programs, and create mechanisms for conflict resolution.

Examples of successful international cooperation in transboundary river basins include the Nile Basin Initiative and the Mekong River Commission. The Nile Basin Initiative, involving 10 African countries, seeks to promote equitable water use, economic development, and environmental sustainability within the Nile River Basin. Similarly, the Mekong River Commission, comprising

Cambodia, Laos, Thailand, and Vietnam, works to manage the Mekong River's resources through joint research, planning, and development. These initiatives highlight the importance of international cooperation in managing transboundary water resources and provide valuable lessons for other regions facing similar challenges.

Water governance frameworks provide the structure necessary to manage resources at multiple scales, from local communities to international partnerships. Effective governance at all levels is essential for achieving sustainable water management within river basins, particularly in transboundary settings where cooperation is essential for balancing competing demands.

Institutional Capacity Building

Institutional capacity building is a crucial component of water governance, as it strengthens the abilities of organizations and communities to manage water resources effectively. Capacity building encompasses a wide range of activities, from enhancing the technical skills of water management professionals to fostering leadership within local communities. By building capacity within institutions and among stakeholders, water governance can be made more resilient, adaptive, and responsive to the complex challenges of river basin management.

Strengthening Institutions Responsible for River Basin Management

Institutional capacity building involves strengthening the technical, administrative, and financial capabilities of organizations responsible for managing river basins. Effective water governance requires institutions that are well-equipped to handle data collection, policy implementation, stakeholder engagement, and conflict resolution. Capacity-building efforts may include providing training on the latest water management technologies, offering workshops on policy analysis and development, and increasing access to financial resources for infrastructure development and maintenance. By enhancing the capacities of these institutions, water governance frameworks can operate more efficiently, making informed decisions that support sustainable resource management.

Capacity Building for Local Communities and Stakeholders

Local communities and stakeholders play a critical role in water governance, particularly in river basins where resource management often relies on community-based initiatives. Capacity building at the community level involves equipping stakeholders with the knowledge, skills, and resources needed to participate meaningfully in governance processes. This may include training in water conservation techniques, education on the ecological impacts of resource use, and the development of community-based monitoring programs. Empowering local stakeholders to engage in governance not only fosters a sense of ownership and accountability but also enables communities to contribute valuable insights and local knowledge to decision-making processes. By building capacity at both institutional and community levels, water governance frameworks become more inclusive and resilient, with a stronger foundation for addressing the complex challenges of river basin management.

Institutional capacity building enhances the ability of organizations and communities to manage water resources sustainably. Strengthening institutions and empowering local stakeholders are essential steps in creating a robust governance framework that supports long-term water resource sustainability within river basins.

Case Studies of Effective Water Governance

Case studies provide valuable insights into the practical application of governance frameworks within river basins. By examining successful governance initiatives, we can identify best practices and lessons that contribute to sustainable river basin management.

The Murray-Darling Basin, Australia

The Murray-Darling Basin Authority (MDBA) is a prime example of effective water governance in action. The MDBA was established to oversee the sustainable management of water resources across Australia's largest river system, which spans four states and supports significant agricultural activity. The MDBA's governance framework includes extensive stakeholder engagement, transparent decision-making processes, and a commitment to balancing water allocations between human uses and ecological needs. By implementing a cap on water extractions, enforcing strict water quality standards, and investing in infrastructure improvements, the MDBA has worked to protect the health of the basin's ecosystems while supporting the needs of local communities and industries.

The Colorado River Basin, United States and Mexico

The governance of the Colorado River Basin illustrates the importance of international cooperation in transboundary water management. The Colorado River, which flows through seven U.S. states and into Mexico, has been subject to complex water-sharing agreements to ensure equitable distribution of resources. The 1944 U.S.-Mexico Water Treaty and subsequent agreements, such as Minute 319 and Minute 323, have established a cooperative framework for managing the basin's resources. These agreements allow for joint monitoring, data sharing, and coordinated conservation efforts. By fostering collaboration between the U.S. and Mexico, the governance framework of the Colorado River Basin has improved water security, enhanced ecosystem health, and strengthened diplomatic relations.

The Rhine River Basin, Europe

The Rhine River Basin provides an example of successful water governance through international cooperation. The International Commission for the Protection of the Rhine (ICPR), involving six European countries, has worked to address pollution, flood risks, and biodiversity loss within the basin. Through joint monitoring programs, harmonized policies, and collaborative restoration efforts, the ICPR has achieved significant improvements in water quality and ecosystem health. The success of the Rhine governance model demonstrates the importance of cooperation, transparency, and shared responsibility in managing transboundary water resources.

These case studies highlight the effectiveness of governance frameworks that prioritize collaboration, transparency, and stakeholder engagement. By examining successful governance initiatives, we can identify best practices and lessons that contribute to sustainable river basin management, providing a roadmap for other regions facing similar challenges.

Conclusion

Water governance plays a critical role in sustainable river basin management by providing the structures, policies, and processes required to allocate resources, resolve conflicts, and protect ecosystems. Good governance ensures that decisions are made transparently, equitably, and in a way that considers the long-term needs of both human and ecological systems. By addressing challenges such as fragmented responsibilities and inadequate policies, governance frameworks can support sustainable resource management across multiple scales, from local communities to international partnerships.

Governance frameworks operate at different levels, including national, regional, and local structures, each with a unique role in water management. Effective governance in transboundary river basins often requires international cooperation, as seen in initiatives like the Nile Basin Initiative and the Mekong River Commission. Institutional capacity building further strengthens water governance by enhancing the capabilities of organizations and empowering local communities to participate in decision-making. Finally, case studies of successful governance initiatives, such as those in the Murray-

Darling, Colorado, and Rhine river basins, provide valuable insights into best practices and lessons learned in water governance.

In conclusion, effective water governance is essential for achieving sustainable river basin management. By building strong governance frameworks, fostering cooperation across boundaries, and empowering stakeholders, water managers can create a resilient foundation for the sustainable use and protection of river basins in an increasingly challenging environmental landscape.

Chapter 4: Ecosystem-Based Approaches to River Basin Management

Ecosystem-based approaches to river basin management recognize that healthy ecosystems are crucial to the resilience, productivity, and sustainability of water resources. As demands on river basins intensify due to urban expansion, industrial development, agricultural practices, and climate change, conventional management methods often fail to address the full complexity of river systems. Rather than solely relying on artificial infrastructure or traditional engineering solutions, ecosystem-based management incorporates the functions and services provided by nature itself. This includes conserving, restoring, and enhancing ecosystems such as wetlands, forests, and floodplains, which naturally regulate water flows, maintain water quality, support biodiversity, and protect communities from extreme weather events. Ecosystem-based approaches emphasize these natural functions, aiming to sustain both human and ecological needs within the river basin.

In this chapter, we will first explore the essential roles ecosystems play in maintaining river basin sustainability, detailing the contributions of wetlands, forests, and floodplains as natural water regulators. Following this, we discuss Ecosystem-Based Management (EBM), a comprehensive strategy that balances environmental, social, and economic needs. Finally, we examine nature-based solutions (NbS) for river basin management, covering practices like green infrastructure, reforestation, and wetland restoration, which enhance ecosystem services to meet water management goals. By incorporating ecosystem-based approaches, river basin managers can create adaptable and resilient frameworks that benefit both people and the environment.

The Role of Ecosystems in River Basin Sustainability

The ecosystems within a river basin form an interconnected web of habitats and natural functions that support water quality, regulate flows, and provide critical ecosystem services. Recognizing and preserving these functions is essential for the long-term sustainability of river basins.

Wetlands, Forests, and Floodplains as Natural Water Regulators

Each ecosystem within a river basin contributes unique services that enhance water regulation and support biodiversity. Wetlands, for example, function as natural water filters, trapping pollutants, sediments, and excess nutrients from surface runoff, particularly from agricultural areas. This process improves downstream water quality, benefiting aquatic ecosystems and reducing the need for costly water treatment infrastructure. Wetlands also play a critical role in flood control, as they absorb and retain excess water during heavy rainfall events, releasing it gradually back into rivers, which reduces peak flows and helps prevent downstream flooding.

Forests, especially those located in upper watersheds, are vital for maintaining soil stability and water quality. Trees and vegetation prevent soil erosion by holding soil in place with their root systems, reducing sedimentation in rivers and reservoirs. Forests also play a key role in the water cycle through the process of transpiration, in which water moves from the soil through vegetation and into the atmosphere, maintaining humidity levels and supporting precipitation patterns. During dry periods, forested areas can help sustain base flows in rivers, supporting both aquatic ecosystems and human water needs.

Floodplains, which are low-lying areas adjacent to rivers, provide critical ecosystem services that buffer river systems from extreme fluctuations in water levels. During high-flow events, floodplains act as natural storage areas, absorbing excess water and releasing it gradually, which reduces the risk of downstream flooding. As water flows across floodplains, it slows down, allowing sediment and nutrients to settle. This process enriches soil fertility, benefiting agricultural productivity. Floodplains also serve as important habitats for diverse species and as migratory corridors for fish and wildlife, enhancing biodiversity within the river basin.

Importance of Preserving Ecosystem Services

The services provided by wetlands, forests, and floodplains are invaluable for maintaining river basin health and resilience. These ecosystem functions are particularly critical in the face of climate change, which is expected to increase the frequency and intensity of extreme weather events, such as floods and droughts. By maintaining and restoring natural habitats within river basins, managers can harness the resilience of ecosystems to buffer against environmental stressors. Moreover, preserving ecosystem services can reduce the need for expensive artificial infrastructure. For instance, a well-maintained wetland system can often filter water more effectively and sustainably than conventional treatment plants, highlighting the economic and ecological benefits of conserving natural ecosystems.

The role of ecosystems in river basin sustainability highlights the need for an integrated approach that values natural processes and functions. By recognizing the benefits of ecosystem services and prioritizing their protection, river basin managers can support both ecological integrity and human well-being.

Ecosystem-Based Management

EBM is a holistic approach that aims to sustain and enhance the health of ecosystems by managing human activities within their natural limits. EBM in river basins integrates ecological, social, and economic objectives, acknowledging that the well-being of human populations is deeply interconnected with ecological health.

Concept and Principles of EBM in River Basin Contexts

The foundational principle of EBM is that ecosystems, rather than individual resources, are the primary units of management. EBM considers all physical, chemical, and biological components of an ecosystem and their interactions with human activities. The approach emphasizes adaptive management, precautionary measures, and the integration of multiple objectives to support sustainable resource use.

In the context of river basin management, EBM seeks to balance water use for human purposes—such as drinking water supply, irrigation, and industrial processes—with the needs of ecosystems. This balance is achieved by setting environmental flow standards, protecting critical habitats, and minimizing disruptions to natural hydrological processes. EBM also involves managing land use within the basin, as land-cover changes, such as deforestation or urbanization, can have significant impacts on water quality, sedimentation, and habitat loss. By prioritizing ecosystem health, EBM supports the resilience of river basins, enabling them to adapt to environmental changes and recover from disturbances.

Tools and Methodologies for Implementing EBM

Implementing EBM requires a variety of tools and methodologies that support data-driven, adaptive decision-making. One foundational tool is ecological monitoring, which involves tracking indicators of ecosystem health, such as water quality, species diversity, and habitat conditions. Monitoring allows managers to assess the effectiveness of EBM measures and make adjustments as needed to address emerging threats or changes in the ecosystem.

Geographic Information Systems (GIS) and remote sensing are also essential tools for EBM, providing spatial data on land use, vegetation cover, and hydrological patterns within a river basin. These technologies help managers identify critical habitats, assess land-use impacts, and monitor changes in ecosystem conditions over time. Hydrological modeling is another important tool, allowing managers to simulate water flows under different scenarios and assess the potential impacts of management decisions on water availability and ecosystem health.

Stakeholder engagement is crucial for implementing EBM, as it fosters collaboration and builds support for conservation initiatives.

By involving local communities, industries, and other stakeholders in the planning process, EBM can incorporate diverse perspectives and address potential conflicts between human needs and ecological goals. Public awareness campaigns, community-led monitoring programs, and participatory planning workshops are all effective methods for engaging stakeholders in EBM.

EMB provides a framework for managing river basins in ways that prioritize ecological health and sustainability. By using a combination of monitoring, modeling, and stakeholder engagement, EBM supports adaptive, data-driven approaches that respond to the complexities of river basin ecosystems.

Nature-Based Solutions for River Basin Management

NbS refer to actions that work with and enhance natural processes to address environmental challenges. In river basin management, NbS offer practical and cost-effective approaches to improving water quality, reducing flood risks, and enhancing ecosystem resilience. These solutions include practices like green infrastructure, reforestation, and wetland restoration, which leverage the natural functions of ecosystems to provide essential services.

Green Infrastructure, Reforestation, and Wetland Restoration

Green infrastructure involves using natural or semi-natural systems to manage water resources. Examples of green infrastructure in river basins include riparian buffers, permeable surfaces, and rain gardens, which help filter pollutants, reduce runoff, and improve water infiltration. Riparian buffers—strips of vegetation planted along riverbanks—are especially effective for controlling sedimentation and nutrient loading, as they trap sediments and absorb excess nutrients before they enter waterways.

Reforestation, or the planting of trees and vegetation in degraded or deforested areas, enhances river basin health by stabilizing soils, reducing erosion, and improving groundwater recharge. Trees slow the movement of water through the landscape, supporting water infiltration and reducing surface runoff. Reforestation also supports biodiversity by creating habitats for a wide range of species and increasing connectivity within the basin.

Wetland restoration involves rehabilitating degraded wetlands or creating new wetland areas to restore their natural functions. Restored wetlands act as natural water filters, absorbing pollutants and nutrients while providing habitats for numerous species. Wetlands also store large volumes of water, mitigating flood risks by absorbing peak flows during heavy rainfall events. Wetland restoration is particularly beneficial in urbanized or agricultural regions, where wetland loss has reduced the landscape's capacity to manage water and support biodiversity.

Benefits of Integrating Nature-Based Solutions in Basin Management

Incorporating nature-based solutions into river basin management offers numerous environmental and economic benefits. One primary advantage of NbS is their ability to enhance ecosystem resilience, making river basins more capable of withstanding extreme weather events, pollution, and human impacts. For instance, a river basin with well-maintained forests and wetlands is better equipped to handle flood events, as these ecosystems can absorb and store large amounts of water, reducing impacts on downstream communities.

Nature-based solutions also provide cost-effective alternatives to artificial infrastructure. Unlike traditional solutions, such as dams and levees, which require regular maintenance and eventual replacement, NbS are often self-sustaining once established. By investing in green infrastructure, reforestation, and wetland restoration, river basin managers can achieve long-term water management goals while saving on costs associated with engineered solutions.

Moreover, NbS offer social and cultural benefits by creating green spaces, enhancing recreational opportunities, and preserving natural landscapes that hold cultural significance. These aspects contribute to community well-being, fostering a sense of connection to the environment and encouraging stewardship of natural resources. By integrating NbS into river basin management, managers address ecological and economic needs while promoting social resilience and community engagement.

Nature-based solutions represent a flexible and adaptive approach to river basin management, providing a practical strategy that aligns with the principles of Ecosystem-Based Management. By leveraging natural systems, NbS offer resilient solutions to water management challenges that benefit both ecosystems and human populations.

Conclusion

Ecosystem-based approaches to river basin management emphasize the role of healthy ecosystems in maintaining water quality, regulating flows, and supporting biodiversity. By preserving and restoring key habitats within a river basin, managers can harness the resilience of natural systems to buffer against environmental stressors, such as floods, droughts, and pollution. Wetlands, forests, and floodplains serve as natural infrastructure, providing services that enhance the sustainability and resilience of river basins.

EBM offers a framework that integrates ecological health with human needs, employing a range of tools and methodologies, such as ecological monitoring, GIS, and stakeholder engagement. Additionally, nature-based solutions, including green infrastructure, reforestation, and wetland restoration, offer cost-effective, resilient alternatives to conventional engineering projects. By incorporating these solutions, river basin managers can enhance ecosystem resilience, reduce flood risks, and improve water quality, achieving outcomes that support both people and nature.

In conclusion, ecosystem-based approaches provide a sustainable pathway for river basin management that respects and harnesses the power of natural systems. By integrating EBM principles and nature-based solutions, river basin managers can develop adaptive, resilient, and ecologically sound strategies that support the well-being of communities and the health of river basin ecosystems.

Chapter 5: Climate Change and River Basin Resilience

The impacts of climate change are increasingly visible and transformative across global river basins, creating profound challenges for sustainable water management. River basins, which encompass diverse ecosystems and human communities, are especially vulnerable to changing climate patterns. Variations in precipitation, increased frequency of extreme weather events, and altered water availability are all part of the new challenges that river basin managers must consider. The resilience of river basins to these changes is essential to ensure the sustainability of water resources, support biodiversity, and protect communities dependent on these ecosystems for their livelihoods. Building resilient river basins that can adapt to climate-driven changes requires forward-thinking management strategies that incorporate adaptation measures and proactive planning.

In this chapter, we will examine the specific impacts of climate change on river basins, including altered precipitation patterns, the increased occurrence of extreme weather events, and changes in water availability. Next, we discuss adaptive management strategies designed to build resilient river basins, exploring how these approaches can enhance the capacity of river basins to withstand and adapt to climate-related stressors. Finally, we will present case studies highlighting successful climate adaptation strategies in river basin management, providing practical examples of how resilience can be achieved through thoughtful planning and innovation.

Climate Change Impacts on River Basins

Climate change has far-reaching effects on river basins, altering hydrological patterns, increasing the frequency of extreme weather events, and intensifying the vulnerabilities of ecosystems and human communities within these basins. Understanding these impacts is essential for developing strategies to build resilience and adapt to future conditions.

Altered Precipitation Patterns, Extreme Weather Events, and Water Availability

One of the most significant impacts of climate change on river basins is the alteration of precipitation patterns. In many regions, climate change has led to shifts in seasonal rainfall, with some areas experiencing more intense rainfall during shorter periods and others experiencing prolonged dry spells. These changes affect the overall availability of water within river basins, with some regions facing reduced water levels and others dealing with increased flooding risks. Altered precipitation patterns can disrupt natural flow regimes, impacting both aquatic ecosystems and human water needs.

Extreme weather events, such as intense storms, hurricanes, and heatwaves, are also becoming more common due to climate change. These events exacerbate the stress on river basins by causing sudden surges in water flow that can lead to flooding, erosion, and sedimentation. The increased frequency and intensity of these events pose significant challenges to river basin infrastructure, as dams, levees, and water treatment plants may not be designed to handle such extreme conditions. Prolonged heatwaves, combined with reduced rainfall, can exacerbate drought conditions, placing additional pressure on water resources within the basin.

Changes in water availability, influenced by both altered precipitation and increased evapotranspiration due to rising temperatures, further complicate water management. Reduced water availability during dry seasons affects agriculture, hydropower generation, and drinking water supply, while increased flow variability disrupts the natural processes that support fish spawning, vegetation growth, and wetland health. In snow-fed river basins, such as those in mountainous regions, warming temperatures lead to earlier snowmelt, altering seasonal flow patterns and reducing water availability in late summer, when demand is often highest.

River Basin Vulnerabilities to Climate Change

River basins vary in their vulnerability to climate change, depending on factors such as geography, hydrology, and the resilience of local ecosystems and communities. Mountainous river basins, which rely on glacial meltwater, are particularly vulnerable, as warming temperatures accelerate glacial retreat, reducing the long-term water supply. Glaciers in regions such as the Himalayas, the Andes, and the Alps are shrinking rapidly, posing a severe risk to water availability for downstream communities that depend on consistent meltwater flows.

In arid and semi-arid regions, river basins are highly susceptible to droughts exacerbated by climate change. Prolonged dry spells and increased evapotranspiration rates reduce surface water levels, groundwater recharge, and soil moisture, leading to a scarcity of water for both human and ecological needs. Droughts have far-reaching impacts, including reduced agricultural productivity, heightened competition for water resources, and increased pressure on ecosystems that may already be degraded by human activities.

Flood-prone river basins, particularly those in low-lying or coastal areas, face increased flood risks due to climate-induced sea-level rise and more intense rainfall events. Coastal river basins are also vulnerable to saltwater intrusion, which can contaminate freshwater supplies and harm agricultural lands. Additionally, river basins in tropical and subtropical regions may experience shifts in disease dynamics, as warmer temperatures and altered water conditions create favorable environments for waterborne diseases and vectors, such as mosquitoes.

The impacts of climate change on river basins highlight the urgent need for adaptive management strategies that can enhance resilience and protect ecosystems and communities from climate-related threats.

Building Resilient River Basins

Building resilience in river basins requires a proactive approach that incorporates adaptive management practices and climate change adaptation strategies. By fostering flexibility and responsiveness in water management, these approaches can help river basins withstand and recover from climate-related disruptions.

Adaptive Management Approaches to Increase Resilience

Adaptive management is a dynamic approach that allows river basin managers to adjust strategies and practices in response to changing conditions and new information. Rather than following rigid management plans, adaptive management involves continuous monitoring, evaluation, and adjustment, enabling managers to respond effectively to climate variability and unforeseen challenges. In the context of climate change, adaptive management is particularly valuable, as it provides a framework for dealing with uncertainties and incorporating new scientific insights into decision-making.

Key components of adaptive management include monitoring environmental indicators, evaluating the effectiveness of interventions, and making iterative adjustments based on observed outcomes. For example, if a river basin experiences a prolonged drought, adaptive management allows for adjustments in water allocation policies, prioritizing essential uses and implementing conservation measures to reduce demand. Adaptive management also facilitates collaboration among stakeholders, as it encourages regular communication and joint problem-solving, fostering a shared commitment to building resilience.

Adaptive management practices can be applied across various sectors within the river basin, including agriculture, energy, and urban water supply. In agriculture, for instance, adaptive management may involve promoting drought-resistant crops, improving irrigation efficiency, and diversifying water sources. In the energy sector, it may involve adjusting hydropower operations to accommodate fluctuations in water availability. By embracing adaptive management, river basin managers can create a flexible, resilient framework that can accommodate both current and future climate challenges.

Incorporating Climate Change Adaptation into River Basin Planning

Integrating climate change adaptation into river basin planning involves designing management strategies that explicitly account for the anticipated impacts of climate change. This process begins with a comprehensive vulnerability assessment, which identifies the specific risks and vulnerabilities faced by the river basin due to climate change. Vulnerability assessments consider factors such as water availability, ecosystem health, infrastructure resilience, and community needs, providing a basis for targeted adaptation measures.

Once vulnerabilities are identified, river basin managers can develop adaptation strategies that address these specific risks. Strategies may include enhancing water storage capacity, restoring degraded ecosystems, and implementing conservation practices to reduce water demand. Climate adaptation planning also involves incorporating climate projections into water management models, allowing managers to simulate different scenarios and assess the potential impacts of various adaptation measures. By using climate models, river basin managers can make informed decisions about infrastructure investments, water allocation policies, and ecosystem conservation efforts, ensuring that management plans are robust enough to withstand future climate conditions.

Effective climate adaptation planning also requires stakeholder engagement, as the success of adaptation measures often depends on the support and cooperation of local communities, industries, and government agencies. Engaging stakeholders in the planning process ensures that adaptation strategies are socially acceptable and economically viable, increasing the likelihood of successful implementation. Additionally, stakeholder involvement fosters a sense of shared responsibility for resilience, as communities become active participants in managing climate risks within the river basin.

Building resilient river basins through adaptive management and climate adaptation planning provides a proactive approach to addressing the impacts of climate change. By integrating these strategies into river basin management, managers can enhance the resilience of both ecosystems and human communities, creating a sustainable framework that supports long-term water security.

Case Studies on Climate Adaptation in River Basins

Case studies of climate adaptation in river basins offer valuable insights into practical approaches for building resilience and protecting water resources. These examples demonstrate successful strategies for adapting to climate change and provide lessons that can inform future river basin management efforts.

The Murray-Darling Basin, Australia

The Murray-Darling Basin in Australia provides an example of adaptive management in response to climate variability and prolonged droughts. As one of Australia's most significant agricultural regions, the Murray-Darling Basin has faced severe water scarcity due to climate-induced reductions in rainfall and increased evaporation rates. In response, the Murray-Darling Basin Authority (MDBA) has implemented a range of adaptive management strategies to enhance resilience and reduce water demand.

One key adaptation measure has been the establishment of water allocation plans that prioritize essential uses, such as drinking water and ecosystem health, over agricultural and industrial uses during periods of drought. Additionally, the MDBA has promoted water-efficient irrigation practices, supported the use of drought-resistant crops, and implemented water trading mechanisms that allow farmers to buy and sell water rights, encouraging more efficient use of limited resources. Through these adaptive strategies, the Murray-Darling Basin has increased its resilience to climate variability, supporting both agricultural productivity and ecological health.

The Colorado River Basin, United States and Mexico

The Colorado River Basin, which supplies water to seven U.S. states and Mexico, faces significant climate challenges due to reduced snowpack, prolonged droughts, and increased water demand. In response, the U.S. and Mexico have developed a cooperative adaptation strategy that includes water-sharing agreements, conservation initiatives, and ecosystem restoration efforts.

Under the Minute 323 agreement, the U.S. and Mexico have committed to jointly managing the Colorado River's resources, with a focus on improving water conservation and enhancing resilience to climate change. This agreement includes measures such as increasing water storage capacity, restoring wetlands along the river, and implementing water-saving technologies. By fostering cross-border cooperation, the Colorado River Basin adaptation strategy demonstrates the importance of collaborative approaches in managing shared water resources under changing climate conditions.

The Rhine River Basin, Europe

The Rhine River Basin, which spans six European countries, has implemented a climate adaptation strategy that focuses on flood management and ecosystem restoration. Due to its location in a temperate region, the Rhine Basin is vulnerable to increased rainfall and flooding, exacerbated by climate change. The International Commission for the Protection of the Rhine (ICPR) has developed an adaptation plan that includes floodplain restoration, flood forecasting, and the construction of flood control infrastructure.

One of the key adaptation measures has been the restoration of natural floodplains, which provides additional storage capacity for floodwaters and reduces the risk of downstream flooding.

Additionally, the ICPR has implemented a comprehensive flood forecasting system, allowing communities along the Rhine to prepare for and respond to flood events more effectively. By combining ecosystem restoration with technological solutions, the Rhine Basin adaptation strategy enhances resilience to climate-induced flooding and protects both human communities and natural habitats.

These case studies illustrate the diverse strategies that can be employed to build climate resilience in river basins. By adopting adaptive management practices, fostering cross-border cooperation, and restoring ecosystems, river basin managers can create robust frameworks that support sustainable water management under changing climate conditions.

Conclusion

Climate change presents complex and significant challenges for river basin management, as altered precipitation patterns, extreme weather events, and shifts in water availability threaten the resilience of ecosystems and communities. Building resilient river basins that can adapt to these changes requires a proactive approach that incorporates adaptive management practices, climate adaptation planning, and collaboration among stakeholders. Adaptive management provides a flexible framework that allows river basin managers to adjust strategies in response to changing conditions, while climate adaptation planning ensures that management approaches are robust enough to withstand future climate impacts.

The case studies of the Murray-Darling, Colorado, and Rhine river basins offer valuable lessons on successful climate adaptation strategies, highlighting the importance of cooperation, conservation, and innovative management practices. By learning from these examples, river basin managers can develop resilience strategies that are tailored to the specific vulnerabilities and needs of their regions, supporting sustainable water management in the face of climate change.

In conclusion, addressing climate change in river basin management requires a commitment to resilience, adaptation, and innovation. By integrating climate adaptation measures into river basin planning, managers can protect water resources, support biodiversity, and ensure the long-term sustainability of river basins for future generations.

Chapter 6: Sustainable Water Use Across Sectors

Water is an essential resource for human well-being, economic prosperity, and ecological health. However, increasing water demands from agriculture, industry, and urban populations place significant pressure on river basins. Unsustainable water use not only depletes resources but also degrades ecosystems, reduces water quality, and threatens the livelihoods of communities dependent on river basins. Balancing water needs across different sectors is essential for ensuring long-term sustainability, especially as climate change, population growth, and industrial development intensify demands on water resources. Achieving sustainable water use requires sector-specific strategies that focus on efficient water use, pollution reduction, and conservation practices tailored to the unique needs of agriculture, industry, and urban environments.

This chapter explores how sustainable water use can be achieved across key sectors, starting with agriculture, the largest global consumer of water. We will examine how improving irrigation efficiency, adopting water-saving technologies, and promoting sustainable agricultural practices can significantly reduce the sector's water footprint. We then analyze water use in industrial sectors, focusing on strategies to minimize consumption and limit environmental impact, particularly in high-demand industries such as mining, energy, and manufacturing. Finally, we turn to urban water use, where the rapid growth of cities requires innovative solutions to manage water demand sustainably, including water recycling, rainwater harvesting, and green infrastructure. Each of these sectors plays a vital role in supporting sustainable water use within river basins.

Water Use in Agriculture

Agriculture accounts for approximately 70% of global freshwater withdrawals, making it the most significant contributor to water consumption within river basins. The heavy reliance on water for crop irrigation, livestock, and food processing has a profound impact on water availability, quality, and ecosystem health in many regions. Addressing the challenges of agricultural water use is essential for promoting sustainability within river basins and ensuring food security for growing populations.

Irrigation Practices, Water Efficiency, and Impact on River Basins

Irrigation is central to modern agriculture, allowing crops to thrive in regions with limited rainfall and supporting intensive farming practices. However, traditional irrigation methods, such as flood irrigation, often waste large amounts of water through evaporation, runoff, and percolation. These inefficiencies place considerable stress on river basins, especially during periods of low rainfall or drought, when water levels are naturally reduced. In regions dependent on irrigation, excessive water extraction can deplete rivers, lakes, and aquifers, harming both ecosystems and communities that rely on these resources for drinking water, sanitation, and economic activities.

To reduce the water demands of agriculture, improving irrigation efficiency is a primary strategy for sustainable water use. Efficient irrigation systems, such as drip and sprinkler irrigation, deliver water directly to the plant roots, minimizing water losses and ensuring that crops receive the right amount of water. Drip irrigation, for example, reduces water use by 30–70% compared to traditional flood irrigation, while also improving crop yields. Precision agriculture, which uses sensors and data analytics to monitor soil moisture and adjust water application based on real-time data, further enhances water efficiency. Precision irrigation ensures that water is applied only when and where it is needed, reducing waste and maximizing crop productivity.

Sustainable Agriculture and Water-Saving Technologies

Sustainable agriculture aims to balance productivity with environmental stewardship, promoting practices that conserve water, reduce pollution, and support soil health. Water-saving technologies, such as soil moisture sensors, automated irrigation systems, and drought-resistant crop varieties, are essential tools for achieving sustainable water use in agriculture. Soil moisture sensors provide farmers with real-time data on soil water levels, enabling them to make informed decisions about irrigation scheduling and avoid overwatering. Automated irrigation systems can be programmed to adjust water application based on weather forecasts, soil moisture, and crop needs, further improving water efficiency and reducing stress on river basins.

In addition to technological solutions, sustainable agriculture promotes practices that enhance soil health and water retention, reducing the need for irrigation. Conservation tillage, for example, involves minimal soil disturbance, preserving soil structure and reducing erosion. Healthy soils retain more moisture, allowing crops to withstand dry conditions and reducing the frequency of irrigation. Cover cropping and crop rotation are other sustainable practices that improve soil health, enhance water retention, and reduce nutrient runoff. By integrating these practices, farmers can reduce water demand, improve resilience to climate variability, and protect river basin resources.

The development of drought-resistant crop varieties is another important aspect of sustainable agriculture, particularly in arid and semi-arid regions. These crop varieties are bred to require less water and are more tolerant to dry conditions, reducing the water footprint of agriculture and enabling farmers to maintain productivity during periods of water scarcity. By adopting water-saving technologies and sustainable practices, the agricultural sector can significantly reduce its water footprint, alleviate pressure on river basins, and contribute to the overall sustainability of water resources.

Sustainable water use in agriculture is critical for preserving river basin health and ensuring food security. By implementing efficient irrigation practices, adopting water-saving technologies, and promoting sustainable farming methods, agriculture can meet the growing demand for food while protecting water resources within river basins.

Industrial Water Use

The industrial sector is a significant consumer of water, especially in water-intensive industries such as mining, energy, and manufacturing. Industrial water use not only places a strain on water resources but also contributes to pollution, as many industrial processes generate wastewater containing chemicals, heavy metals, and other contaminants. Sustainable water management in industry is essential to reduce water demand, protect river basins, and minimize environmental impact.

Water-Intensive Industries: Mining, Energy, and Manufacturing

Some industries, such as mining, energy production, and manufacturing, require substantial quantities of water for their operations. Mining, for instance, uses water in various stages of mineral extraction, processing, and dust suppression. The energy sector, particularly in thermal power generation, relies on water for cooling purposes, with power plants withdrawing large amounts of water from nearby rivers or reservoirs. Hydropower, a renewable energy source, also requires consistent river flows to generate electricity, directly impacting water availability within river basins. Manufacturing industries, including textiles, chemicals, and food processing, are similarly dependent on water for cleaning, cooling, and processing raw materials, contributing to high water demand in industrial regions.

The water consumption of these industries can reduce water availability for other uses, impacting local communities, agriculture, and ecosystems. Moreover, industrial activities often produce pollutants, such as heavy metals, dyes, and organic compounds, which contaminate water sources and degrade water quality. Industrial wastewater, if not properly treated, poses serious risks to human health, aquatic life, and soil fertility, making pollution control a critical component of sustainable industrial water use.

Strategies for Reducing Industrial Water Consumption and Pollution

Reducing industrial water use and pollution requires a combination of technological innovations, regulatory frameworks, and best practices. Water recycling and reuse are effective strategies for reducing water demand, as they allow industries to treat and reuse wastewater within their operations, minimizing the need for freshwater withdrawals. Closed-loop systems, for example, capture and treat wastewater on-site, enabling its reuse in processes such as cooling, washing, and product rinsing. This reduces the overall water footprint of industrial facilities and alleviates pressure on river basins.

Advanced treatment technologies, such as membrane filtration, reverse osmosis, and chemical precipitation, improve the quality of industrial wastewater by removing contaminants, making it suitable for reuse. In sectors like mining and manufacturing, zero-liquid discharge (ZLD) systems are being adopted to eliminate wastewater discharge altogether. ZLD systems recover water from wastewater through evaporation and condensation, reducing pollution and preserving river basin water quality. These systems are especially useful in regions where water scarcity and pollution pose significant challenges to river basin health.

Regulatory frameworks play a crucial role in promoting sustainable industrial water use by establishing standards for water quality, discharge limits, and pollution controls. Water quality standards and discharge permits incentivize industries to adopt cleaner technologies and minimize their environmental impact. Certification programs, such as ISO 14001 for environmental management, encourage companies to implement sustainable practices and demonstrate their commitment to reducing water use and pollution. By adopting these strategies, industries can reduce their water footprint, protect river basins, and contribute to sustainable water management.

In addition to technological and regulatory solutions, best practices for industrial water management include regular monitoring, employee training, and environmental audits. Regular monitoring of water use and wastewater discharge enables industries to identify areas for improvement and ensure compliance with regulatory standards. Training employees on water conservation practices fosters a culture of sustainability within industrial facilities, encouraging efficient water use and pollution reduction. Environmental audits provide an opportunity for industries to assess their water management practices, identify inefficiencies, and develop targeted solutions for reducing water consumption and pollution.

Sustainable water use in industry is essential for minimizing the environmental impacts of industrial activities on river basins. By implementing water recycling, advanced treatment technologies, regulatory compliance, and best practices, industries can reduce their water consumption and pollution, promoting healthier ecosystems and sustainable water management within river basins.

Urban Water Use

As global urban populations continue to grow, cities face increasing demand for water resources, placing significant pressure on river basins. Urban water use includes residential, commercial, and institutional needs, as well as requirements for public services, green spaces, and infrastructure. Sustainable urban water management is crucial to reduce the strain on river basins, ensure water availability, and create resilient cities that can adapt to climate change and population growth.

Water Demand in Cities and Urban Planning for Sustainable Water Use

Urban areas typically have high water demand due to population density, economic activities, and public services. Residential water use includes drinking, sanitation, and household needs, while commercial and institutional sectors require water for offices, schools, hospitals, and other facilities. Public spaces, such as parks and recreational areas, also require irrigation to maintain green spaces. As cities grow, managing this demand becomes increasingly challenging, particularly in regions with limited water resources.

Urban planning plays a crucial role in promoting sustainable water use, as it provides a framework for integrating water management with land use, transportation, and infrastructure development. By implementing water-sensitive urban design (WSUD) principles, cities can create environments that minimize water consumption, promote water conservation, and protect river basin health. WSUD involves designing urban landscapes to retain and infiltrate rainwater, reducing runoff and increasing groundwater recharge. Green roofs, permeable pavements, and bioswales are examples of WSUD practices that enhance water efficiency and reduce pressure on river basins.

Water Recycling, Rainwater Harvesting, and Green Infrastructure

Sustainable urban water management relies on strategies that reduce water demand and enhance the resilience of cities. Water recycling, for example, involves treating and reusing wastewater for non-potable purposes, such as landscape irrigation, toilet flushing, and industrial processes. This reduces the demand for freshwater and conserves resources within river basins. Some cities have implemented dual-pipe systems, which separate potable and non-potable water, enabling recycled water to be distributed for specific uses.

Rainwater harvesting is another effective strategy, as it captures and stores rainwater from rooftops, pavements, and other surfaces for later use. Harvested rainwater can be used for irrigation, washing, and even potable purposes after proper treatment. Rainwater harvesting reduces the need for mains water, decreases urban runoff, and provides a supplementary water source during dry periods.

Green infrastructure, such as parks, wetlands, and urban forests, supports sustainable water management by enhancing natural water infiltration, reducing stormwater runoff, and providing ecosystem services. Green infrastructure also improves urban resilience to climate change, as it helps manage extreme weather events, such as heavy rainfall and droughts. Urban wetlands, for example, act as natural water storage systems, absorbing excess water during floods and releasing it during dry periods. By incorporating green infrastructure into urban planning, cities can create sustainable landscapes that support water conservation and improve the health of river basins.

Sustainable urban water management is essential for meeting the water needs of growing cities without compromising river basin health. By adopting water recycling, rainwater harvesting, and green infrastructure, urban areas can reduce their water demand, enhance resilience to climate change, and protect the ecosystems that support them.

Conclusion

Sustainable water use across sectors is crucial for preserving the health and resilience of river basins, ensuring that water resources can meet the needs of both human populations and ecosystems. Agriculture, industry, and urban areas are major water consumers, and each sector faces unique challenges in achieving sustainable water management. By implementing sector-specific strategies, these sectors can reduce their water footprint, protect river basin resources, and contribute to a sustainable water future.

In agriculture, improving irrigation efficiency, adopting water-saving technologies, and promoting sustainable farming practices are essential steps to reduce water demand and enhance resilience to climate variability. In the industrial sector, strategies such as water recycling, advanced treatment technologies, and regulatory compliance play a crucial role in minimizing water consumption and pollution, protecting river basin ecosystems. For urban areas, sustainable water management practices, including water-sensitive urban design, rainwater harvesting, and green infrastructure, are vital for managing rising water demand and creating resilient cities.

In conclusion, sustainable water use across sectors is key to achieving long-term water security and maintaining the ecological integrity of river basins. By adopting sustainable practices in agriculture, industry, and urban planning, society can balance economic development with environmental conservation, supporting the health of river basins for future generations.

Chapter 7: Pollution Management in River Basins

River basins, encompassing a complex network of rivers, lakes, wetlands, and groundwater systems, are essential to sustaining biodiversity, human livelihoods, and economic activities. However, pollution from various sources threatens the health and resilience of these water bodies, disrupting ecosystems, degrading water quality, and posing risks to human health. Effective pollution management in river basins is crucial to protect these vital resources and maintain a balance between human use and environmental sustainability.

Pollution within river basins originates from a range of sources, including agricultural runoff, industrial discharges, untreated sewage, and plastic waste. Each type of pollution has unique characteristics and impacts, requiring targeted management approaches that involve monitoring, regulation, prevention, and remediation. In this chapter, we examine the types and sources of pollution that affect river basins, exploring how each contributes to the degradation of water quality and ecosystems. We then discuss strategies for pollution control, including regulatory frameworks, water quality monitoring, and best practices for prevention. Finally, we consider innovative technologies that offer new solutions to address water pollution, including biological approaches and advanced treatment methods. Together, these approaches form the foundation for sustainable pollution management in river basins, protecting both the environment and communities reliant on these waters.

Types and Sources of Pollution in River Basins

Pollution in river basins arises from a diverse array of human activities, each contributing specific types of contaminants that impact water quality and disrupt ecosystems. Understanding these sources and their effects is essential to develop effective pollution management strategies.

Agricultural Runoff

Agriculture is one of the largest sources of pollution in river basins due to the widespread use of fertilizers, pesticides, and animal manure. When it rains, excess fertilizers containing nitrogen and phosphorus wash into nearby rivers and lakes, leading to nutrient pollution. Excess nutrients stimulate the growth of algae, creating algal blooms that deplete oxygen levels in the water when they decompose, creating hypoxic "dead zones" where aquatic life struggles to survive. Additionally, pesticides used in agriculture introduce toxic chemicals into river systems, posing risks to fish, insects, and even humans. Animal manure from livestock operations also contributes to nutrient pollution and introduces pathogens, such as E. coli, that threaten water safety and public health.

Industrial Discharges

Industrial activities, including manufacturing, mining, and chemical production, are major contributors to water pollution in river basins. Many industries discharge wastewater that contains heavy metals, organic chemicals, and other toxic substances, which accumulate in river sediments and enter the food chain. Industries such as mining release metals like mercury, lead, and cadmium, which can have long-lasting and harmful effects on ecosystems and human health. Additionally, industrial wastewater often contains organic pollutants, including solvents, dyes, and oils, which reduce oxygen levels in the water, making it difficult for aquatic organisms to survive. Industrial discharges are a significant concern in regions with limited regulatory oversight, where wastewater treatment infrastructure may be inadequate to prevent pollution.

Sewage and Municipal Wastewater

Untreated or inadequately treated sewage is a prevalent source of pollution in river basins, particularly in rapidly urbanizing areas. Municipal wastewater contains a variety of pollutants, including organic matter, nutrients, pathogens, and pharmaceuticals, all of which affect water quality and ecosystem health. When sewage enters river basins, it introduces bacteria and viruses that pose health risks to humans, especially those who use the river for drinking water or recreation. Additionally, untreated sewage contributes to nutrient pollution, promoting the growth of algae and depleting oxygen levels in the water. In many developing countries, urban areas lack adequate wastewater treatment facilities, leading to significant pollution that impacts both local communities and downstream ecosystems.

Plastic Pollution

Plastic pollution is an emerging issue in river basins, as plastic waste from urban areas, agriculture, and industry accumulates in rivers and lakes. Larger plastic items, such as bottles, bags, and fishing gear, pose physical hazards to wildlife, causing entanglement or ingestion that can be fatal. As plastic waste breaks down, it forms microplastics—small particles that are consumed by fish and other aquatic organisms, entering the food chain and potentially impacting human health. Microplastics are difficult to remove from water and persist in the environment, making plastic pollution a long-term challenge for river basin management. Reducing plastic pollution requires coordinated efforts to prevent plastic waste from entering rivers and to clean up existing debris.

Sedimentation and Erosion

While often overlooked, sedimentation and erosion are significant forms of pollution in river basins, especially in areas with deforestation, agriculture, and construction activities. When vegetation is removed, soil is exposed to rainfall, leading to increased erosion. Sediment-laden runoff from these areas enters rivers, increasing turbidity and reducing water quality. Excessive sedimentation can bury aquatic habitats, suffocate fish eggs, and disrupt the feeding and breeding grounds of many species. Furthermore, sediments often carry attached pollutants, such as pesticides, heavy metals, and nutrients, which contribute to secondary pollution when deposited in river systems.

Each type of pollution has unique effects on river basin ecosystems, degrading water quality, harming aquatic life, and impacting the health and livelihoods of communities. Addressing these diverse sources of pollution is essential to protect river basins and ensure sustainable water management.

Strategies for Pollution Control

Managing pollution in river basins requires a comprehensive approach that includes monitoring, regulatory frameworks, best practices for prevention, and remediation efforts. Effective pollution control strategies are essential to protect water quality and support resilient ecosystems.

Water Quality Monitoring and Regulatory Frameworks

Monitoring water quality is a fundamental aspect of pollution management, as it provides critical data on the health of river basins and helps identify pollution sources. Regular monitoring of parameters such as nutrient levels, dissolved oxygen, pH, and contaminants provides valuable information for assessing pollution trends and evaluating the effectiveness of control measures. Advanced monitoring technologies, such as satellite remote sensing and real-time water quality sensors, enhance the ability to detect pollution events early and respond swiftly. Monitoring data is essential for developing and enforcing regulations, guiding policy decisions, and building public awareness of pollution issues.

Regulatory frameworks establish standards for water quality and set permissible limits for pollutants, ensuring that water quality is maintained at levels that support ecosystem health and public safety. In many regions, regulations such as the U.S. Clean Water Act and the European Union's Water Framework Directive provide legal guidelines for pollution control in river basins. These frameworks include provisions for issuing discharge permits, enforcing compliance, and penalizing violations. Effective regulatory frameworks encourage industries, municipalities, and agricultural operations to adopt pollution control measures. However, in regions with limited enforcement capacity, regulations may be difficult to implement, highlighting the need for institutional strengthening and international support.

Best Practices for Pollution Prevention and Remediation

Preventing pollution at its source is one of the most effective strategies for protecting river basins. In agriculture, best practices include reducing fertilizer and pesticide use, implementing buffer zones along waterways, and promoting soil conservation techniques. Buffer zones are vegetated areas along rivers that filter pollutants from runoff before they reach the water, reducing nutrient and sediment loads. Conservation tillage and cover cropping improve soil health, reduce erosion, and minimize the runoff of agricultural pollutants.

For industrial sectors, pollution prevention measures include wastewater treatment, cleaner production methods, and minimizing the use of hazardous chemicals. Advanced treatment technologies, such as activated carbon filtration, membrane bioreactors, and chemical precipitation, remove contaminants from wastewater before it is discharged into rivers. Industries can also adopt green chemistry principles to design processes that use fewer toxic chemicals, reducing the potential for pollution.

Remediation strategies are essential for addressing legacy pollution in river basins, particularly in areas with accumulated contaminants. Remediation techniques include dredging contaminated sediments, phytoremediation (using plants to absorb pollutants), and bioremediation (using microorganisms to break down contaminants). These methods help restore water quality and rehabilitate ecosystems affected by historical pollution. Although remediation can be costly and time-consuming, it is vital for improving water quality and revitalizing degraded river basins.

Public Awareness and Community Involvement

Educating communities about pollution sources and encouraging public participation in pollution prevention efforts are essential for long-term success. Community-led monitoring programs, where citizens collect water quality data, increase public understanding of pollution issues and promote accountability among local industries and governments. Public awareness campaigns can encourage responsible waste disposal, promote water conservation, and reduce the use of harmful chemicals. Engaging local stakeholders, including farmers, industries, and residents, fosters a shared commitment to pollution control and creates a culture of stewardship within river basins.

Strategies for pollution control, including water quality monitoring, regulatory frameworks, prevention, and remediation, provide essential tools for managing pollution in river basins. By implementing these strategies, river basin managers can protect water quality, support resilient ecosystems, and maintain the health of communities that depend on these vital resources.

Innovative Technologies for Water Pollution Control

Technological advances provide new and effective tools for reducing water pollution and improving pollution management in river basins. Innovative approaches, including biological solutions, advanced treatment methods, and data-driven technologies, offer promising opportunities for addressing complex pollution challenges.

Biological Solutions for Reducing Water Pollution

Biological approaches, such as bioremediation and constructed wetlands, harness natural processes to remove pollutants from water. Bioremediation involves using microorganisms, such as bacteria and fungi, to break down contaminants, such as organic pollutants and heavy metals, in water and sediments. Certain bacteria, for example, can metabolize harmful substances, transforming them into less toxic forms. Bioremediation is particularly effective for treating pollutants that are difficult to remove with conventional methods, such as petroleum hydrocarbons and pesticides. This technique is also relatively low-cost and environmentally friendly, making it an attractive option for pollution control in river basins.

Constructed wetlands are engineered systems that mimic the functions of natural wetlands, using plants, soil, and microorganisms to filter and degrade pollutants. These wetlands are effective at removing nutrients, suspended solids, and pathogens from wastewater, providing a sustainable, low-maintenance solution for improving water quality. Constructed wetlands can be used to treat agricultural runoff, industrial wastewater, and sewage, reducing the pollutant load entering river basins. In addition to water purification, constructed wetlands provide habitat for wildlife, enhance biodiversity, and improve the aesthetic and recreational value of river basins.

Technological Solutions for Reducing Water Pollution

Technological innovations, such as membrane filtration, advanced oxidation processes, and artificial intelligence, offer powerful tools for managing water pollution. Membrane filtration, which uses semi-permeable membranes to separate pollutants from water, is highly effective at removing contaminants, including bacteria, viruses, and heavy metals. This technology is widely used in industrial and municipal wastewater treatment and can produce high-quality water suitable for reuse. Reverse osmosis, a type of membrane filtration, is particularly useful for desalination and treating brackish water, providing a reliable water source for arid regions.

Advanced oxidation processes (AOPs) are chemical treatment methods that use reactive molecules, such as hydroxyl radicals, to break down organic pollutants. AOPs are effective at treating persistent pollutants, such as pharmaceuticals and pesticides, that are resistant to conventional treatment. By breaking down these compounds into harmless byproducts, AOPs improve water quality and reduce the risk of pollution in river basins. While AOPs can be energy-intensive and costly, ongoing research is focused on making these processes more efficient and accessible for widespread use.

Artificial intelligence (AI) and machine learning are emerging as valuable tools for water pollution management, as they enable data-driven decision-making and predictive analysis. AI algorithms can analyze large datasets from water quality sensors, satellite imagery, and weather forecasts, identifying patterns and predicting pollution events. This predictive capability allows river basin managers to take proactive measures, such as adjusting pollution control strategies or issuing early warnings to communities. Machine learning models can also optimize wastewater treatment processes, improving efficiency and reducing operational costs.

Innovative technologies, including biological solutions and advanced treatment methods, provide new avenues for addressing water pollution in river basins. By incorporating these technologies into pollution management efforts, river basin managers can improve water quality, protect ecosystems, and enhance the resilience of river basins to environmental pressures.

Conclusion

Pollution management in river basins is essential for protecting water quality, preserving ecosystems, and ensuring the well-being of communities that depend on these vital resources. Pollution from agriculture, industry, sewage, and plastic waste presents complex challenges that require a multi-faceted approach. By understanding the sources and impacts of pollution, implementing effective control strategies, and leveraging innovative technologies, river basin managers can reduce pollution, improve water quality, and support sustainable water management.

Water quality monitoring and regulatory frameworks provide the foundation for pollution control efforts, ensuring that pollution sources are identified, tracked, and regulated. Best practices for pollution prevention, such as buffer zones, wastewater treatment, and remediation, help reduce pollutant loads and protect aquatic life. Innovative technologies, including bioremediation, constructed wetlands, membrane filtration, and artificial intelligence, offer new tools for tackling persistent and emerging pollutants, enhancing the effectiveness of pollution management efforts.

In conclusion, sustainable pollution management in river basins is crucial for achieving long-term water security, supporting biodiversity, and promoting the health of ecosystems and communities. By adopting a comprehensive approach that combines traditional and innovative solutions, river basin managers can create cleaner, healthier, and more resilient river systems that serve the needs of both people and nature.

Chapter 8: Transboundary River Basin Management

Transboundary river basins, shared by two or more countries, present unique challenges and opportunities in water management. With over 260 transboundary river basins globally, these shared water resources are essential for millions of people who depend on them for drinking water, agriculture, energy, and ecosystem services. However, managing these basins is often complicated by conflicting interests, political tensions, and legal challenges. Upstream nations may prioritize water use for domestic needs, while downstream nations rely on consistent river flows for their own populations, leading to disputes over water allocation, quality, and use. Addressing these issues requires effective governance frameworks, international cooperation, and innovative strategies that foster sustainable, equitable management of shared water resources.

In this chapter, we explore the challenges associated with managing transboundary river basins, focusing on the conflicting interests of upstream and downstream countries and the legal and political complexities involved. We then examine international agreements and case studies that demonstrate effective cooperation, highlighting the role of international organizations in facilitating collaboration. Finally, we discuss strategies for enhancing transboundary cooperation, including conflict resolution mechanisms, benefit-sharing arrangements, and cooperative water management practices. Together, these approaches offer pathways to achieve sustainable transboundary river basin management that supports both ecological integrity and regional stability.

Challenges of Managing Transboundary River Basins

Transboundary river basins are subject to a range of challenges that complicate sustainable water management. The conflicting interests of upstream and downstream nations, along with the legal and political complexities involved, make it difficult to reach agreements that satisfy all parties.

Conflicting Interests Between Upstream and Downstream Nations

In transboundary river basins, upstream and downstream countries often have differing priorities and interests regarding water use, leading to competition and potential conflict. Upstream nations, having control over the source of the river, may wish to use water resources for agriculture, hydropower, or industrial development, which can impact water availability and quality downstream. For example, if an upstream country constructs a dam or diverts water for irrigation, downstream countries may experience reduced water flows, affecting their ability to meet agricultural, domestic, and environmental needs.

These conflicting interests create a power imbalance, as upstream nations have physical control over the flow of water. This control can lead to tensions, particularly during periods of drought or when water demands increase due to population growth or economic development. In many cases, downstream countries are left vulnerable, with limited recourse to influence upstream water use decisions. This imbalance is particularly evident in basins like the Nile, where Egypt, a downstream nation, relies heavily on river flows originating in upstream countries, such as Ethiopia. Egypt's dependence on the Nile for agriculture and drinking water creates a source of tension with Ethiopia, which has undertaken dam projects to meet its own water and energy needs.

Legal and Political Complexities in Transboundary Water Management

Managing shared river basins involves navigating complex legal and political landscapes, as countries often have their own policies, priorities, and legal frameworks for water management. Transboundary water management requires cooperation across political boundaries, yet differing legal systems, environmental standards, and governance structures complicate this process. In some cases, countries may lack comprehensive water laws, or their legal frameworks may not account for transboundary issues, making it difficult to establish shared rules for water allocation, quality standards, and dispute resolution.

Political tensions between countries can further complicate transboundary water management. Historical conflicts, territorial disputes, and power imbalances can hinder trust and cooperation, making it challenging to negotiate agreements that prioritize shared interests over national interests. Additionally, water management often involves multiple government agencies, each with its own mandate and priorities, which can lead to fragmented and uncoordinated approaches to transboundary water issues. These complexities underscore the need for robust legal frameworks and effective governance structures that facilitate cooperation and support equitable water management in transboundary river basins.

The challenges of managing transboundary river basins highlight the importance of cooperation, transparency, and equitable resource sharing. Addressing these issues requires international agreements, collaborative governance frameworks, and mechanisms for resolving disputes and managing competing interests.

International Agreements and Cooperation

International agreements and cooperation are essential tools for managing transboundary river basins effectively. By establishing formal agreements and fostering collaboration, countries can address water allocation, quality, and dispute resolution issues in ways that support sustainable and equitable water use.

There are numerous examples of successful international agreements that have facilitated cooperation in managing shared river basins. These agreements provide valuable lessons for other regions seeking to improve transboundary water management.

Indus Waters Treaty (India and Pakistan)

The Indus Waters Treaty, signed in 1960 between India and Pakistan, is often cited as one of the most successful examples of transboundary water management. Mediated by the World Bank, the treaty allocates the waters of the Indus River and its tributaries between the two countries, with specific rights and responsibilities for each. Despite political tensions and conflicts between India and Pakistan, the treaty has endured, with both countries adhering to its provisions. The Indus Waters Treaty demonstrates the importance of clear allocation rules, dispute resolution mechanisms, and international mediation in facilitating cooperation between countries with competing interests.

Senegal River Basin (West Africa)

The Senegal River Basin, shared by Senegal, Mauritania, Mali, and Guinea, provides another example of successful transboundary cooperation. In 1972, these countries established the Organisation pour la Mise en Valeur du fleuve Sénégal (OMVS), an organization dedicated to coordinating water management and promoting equitable development in the basin. The OMVS oversees joint projects, such as hydropower generation and irrigation infrastructure, that benefit all member states. Through shared governance and a focus on mutual benefits, the Senegal River Basin countries have managed to balance their interests, demonstrating the effectiveness of benefit-sharing arrangements in promoting cooperation.

Colorado River Compact (United States and Mexico)

The Colorado River Compact, signed in 1922 between the United States and Mexico, allocates the river's water resources between the two countries. Over time, additional agreements, such as the 1944 Water Treaty and the Minute 319 and Minute 323 agreements, have been established to address water allocation, conservation, and environmental concerns. These agreements provide frameworks for managing the river's resources in a way that considers the needs of both countries. Minute 323, for example, focuses on improving water conservation, protecting ecosystems, and enhancing water security for both nations. The Colorado River Compact and its subsequent agreements underscore the importance of adaptability, as the compact has been modified over time to address new challenges, such as drought and climate change.

The Role of International Organizations in Facilitating Cooperation

International organizations play a crucial role in facilitating cooperation and supporting the development of agreements for transboundary river basin management. Organizations like the United Nations (UN), World Bank, and International Union for Conservation of Nature (IUCN) provide technical expertise, financial resources, and neutral platforms for negotiations, helping countries address complex water management challenges.

The UN Convention on the Law of the Non-Navigational Uses of International Watercourses (UN Watercourses Convention), adopted in 1997, provides a legal framework for managing transboundary water resources. The convention emphasizes principles such as equitable and reasonable use, the obligation not to cause significant harm, and cooperation. Although not universally ratified, the convention has influenced the development of regional agreements and serves as a reference for countries negotiating transboundary water management arrangements. Similarly, the UNECE Water Convention, originally established for European countries, promotes cooperation, prevention, control, and reduction of transboundary impacts on water resources, serving as a model for collaborative water management globally.

International organizations also provide funding and support for capacity-building initiatives, enabling countries to develop the technical and institutional capacity needed for effective transboundary water management. By facilitating dialogue, providing resources, and promoting best practices, these organizations help foster cooperation and support the implementation of sustainable water management solutions in transboundary river basins.

International agreements and cooperation play a critical role in managing transboundary river basins, providing frameworks for equitable water allocation, pollution control, and conflict resolution. By building on successful examples and leveraging the support of international organizations, countries can address shared water challenges and promote regional stability and sustainability.

Strategies for Enhancing Transboundary Collaboration

Effective collaboration is essential for managing transboundary river basins sustainably. Strategies such as conflict resolution mechanisms, benefit-sharing, and cooperative water management practices help countries work together to address shared water challenges, promoting equitable and sustainable outcomes.

Conflict Resolution Mechanisms

Conflict resolution mechanisms are vital for managing disputes and fostering cooperation in transboundary river basins. These mechanisms provide structured processes for resolving conflicts over water allocation, quality, and use, helping countries address competing interests constructively. Common conflict resolution mechanisms include negotiation, mediation, and arbitration, each offering different levels of formality and third-party involvement.

Negotiation is often the first step in resolving transboundary water disputes, allowing countries to engage in direct dialogue and seek mutually acceptable solutions. When negotiation reaches an impasse, mediation by a neutral third party can help facilitate dialogue, bridge differences, and identify common ground. International organizations, such as the World Bank and the United Nations, often play a mediating role in transboundary water conflicts, providing expertise and neutral platforms for dialogue.

Arbitration and adjudication provide more formal conflict resolution options, with binding decisions made by an impartial body. These mechanisms are used when countries cannot resolve disputes through negotiation or mediation. For example, the Permanent Court of Arbitration has been involved in resolving disputes over transboundary waters, providing a legal forum for addressing complex issues. While arbitration can be time-consuming and costly, it offers a structured approach to dispute resolution, helping countries manage conflicts and uphold agreements.

Benefit-Sharing and Cooperative Water Management

Benefit-sharing is a collaborative approach that emphasizes the mutual benefits of transboundary water management, rather than focusing solely on water allocation. By identifying shared interests and distributing the benefits of water use equitably, countries can create win-win scenarios that promote cooperation. Benefit-sharing arrangements often involve joint infrastructure projects, such as dams or irrigation systems, that provide economic and social benefits for all parties involved.

For example, in the Senegal River Basin, the OMVS coordinates joint development projects, such as hydropower generation and irrigation, which benefit all member states. By focusing on shared development goals, the OMVS promotes cooperation and reduces tensions between countries. Benefit-sharing arrangements can also involve data sharing, environmental monitoring, and disaster preparedness, enabling countries to manage water resources more effectively and respond to shared challenges, such as drought and flooding.

Cooperative water management frameworks are essential for implementing benefit-sharing arrangements and ensuring that countries work together to achieve sustainable outcomes. These frameworks establish common goals, guidelines, and responsibilities, providing a foundation for coordinated action. Joint water commissions, river basin organizations, and international treaties are examples of cooperative frameworks that enable countries to collaborate on water management. These organizations provide platforms for dialogue, facilitate data sharing, and promote transparency, building trust and fostering long-term cooperation in transboundary river basins.

Strategies for enhancing transboundary collaboration, including conflict resolution mechanisms and benefit-sharing arrangements, provide effective tools for managing shared water resources. By working together to address shared challenges and distribute the benefits of water use equitably, countries can achieve sustainable outcomes that support regional stability and environmental sustainability.

Conclusion

Transboundary river basin management is essential for promoting sustainable water use, protecting ecosystems, and supporting regional stability. Managing shared water resources requires countries to address complex challenges, including conflicting interests, legal and political complexities, and the need for equitable water allocation. International agreements, cooperation, and innovative strategies offer pathways for overcoming these challenges and achieving sustainable transboundary water management.

Successful examples of transboundary agreements, such as the Indus Waters Treaty, the Senegal River Basin Organization, and the Colorado River Compact, demonstrate the potential for cooperation in managing shared rivers. These agreements provide valuable lessons for other regions, highlighting the importance of clear allocation rules, conflict resolution mechanisms, and benefit-sharing arrangements. International organizations, such as the United Nations and the World Bank, play a critical role in facilitating cooperation, providing resources, and supporting the development of legal frameworks that promote equitable water use.

By adopting strategies for enhancing transboundary collaboration, such as conflict resolution mechanisms and cooperative water management practices, countries can address shared water challenges and achieve sustainable outcomes. Transboundary river basin management requires a commitment to collaboration, transparency, and equitable resource sharing, ensuring that water resources can meet the needs of both people and ecosystems for generations to come.

Chapter 9: Stakeholder Engagement in River Basin Management

Effective river basin management is built on the foundation of inclusive and active stakeholder engagement. River basins serve as crucial ecosystems, supporting diverse flora and fauna and providing vital resources for human populations, industries, and agricultural activities. As demands on water resources grow, achieving sustainable management within river basins becomes a complex task, necessitating cooperation across various stakeholder groups with differing perspectives, needs, and responsibilities. Stakeholders, ranging from government agencies and businesses to local communities and non-governmental organizations (NGOs), each play essential roles in shaping water management policies, practices, and goals. Recognizing and engaging these stakeholders in decision-making processes promotes inclusivity, accountability, and mutual trust, all of which are critical to sustainable river basin management.

This chapter delves into the essential role of stakeholder engagement in river basin management, emphasizing why inclusive, collaborative approaches are essential. First, we explore the diversity of stakeholders involved, from government bodies to local communities, and discuss the unique contributions and perspectives they bring to water management. Next, we examine effective methods for stakeholder engagement, including participatory approaches, co-management frameworks, and strategies for fostering trust and collaboration among stakeholders. Finally, we address the challenges inherent in stakeholder engagement, such as power imbalances, conflicting interests, and capacity limitations, offering insights into how these obstacles can be managed. Together, these elements provide a comprehensive understanding of stakeholder engagement, paving the way for more equitable, inclusive, and effective river basin management.

The Role of Stakeholders in River Basin Management

Stakeholder engagement lies at the heart of effective river basin management, as it brings together diverse perspectives that help inform decisions, foster inclusivity, and ensure sustainable outcomes. Recognizing the various types of stakeholders and understanding their specific roles is fundamental to promoting collaboration and ensuring that management practices address the needs and interests of all involved.

Importance of Inclusivity in Decision-Making Processes

Inclusivity is essential to the success of river basin management, as it ensures that the interests, knowledge, and concerns of all stakeholders are taken into account. When stakeholders are actively involved in decision-making, it promotes a sense of ownership, transparency, and trust in the management process. Inclusivity allows for a more holistic understanding of the socio-economic and environmental dynamics within a river basin, as each stakeholder group brings unique insights and experiences. Engaging a broad spectrum of stakeholders also facilitates the early identification and mitigation of potential conflicts, creating a collaborative atmosphere that is conducive to long-term sustainability.

Inclusive decision-making also contributes to resilience by fostering a shared understanding of river basin challenges and enabling stakeholders to develop adaptive strategies. As climate change, economic development, and population growth place new pressures on water resources, having a diverse and well-informed set of stakeholders ensures that management practices remain flexible and responsive. By engaging stakeholders in an inclusive manner, river basin managers can achieve a balance between ecological preservation, social needs, and economic goals.

Types of Stakeholders: Governments, Businesses, NGOs, Local Communities

River basin management encompasses a wide range of stakeholders, each with distinct roles, priorities, and expertise. Understanding these differences is crucial for fostering meaningful collaboration and ensuring that all voices are heard.

- Governments: Government agencies, including national, regional, and local entities, are primary stakeholders in river basin management due to their regulatory authority, policy-making responsibilities, and oversight roles. Governments set the legal and institutional frameworks for water management, enforce regulations, and facilitate coordination among stakeholders. Additionally, governments represent the public interest, ensuring that river basin resources are managed to benefit society as a whole. In many cases, governments act as mediators, balancing competing interests and promoting sustainable practices that protect public health, economic development, and environmental quality.

- Businesses: Businesses, particularly those in water-dependent sectors like agriculture, energy, and manufacturing, are major stakeholders in river basin management. These industries rely on water resources for their operations and have significant impacts on water quality and availability. As a result, businesses have a vested interest in sustainable water management practices that support their long-term viability. Companies can contribute to river basin management by investing in water-efficient technologies, supporting conservation initiatives, and participating in public-private partnerships. Their technical expertise and resources can drive innovative solutions, such as pollution reduction, water recycling, and habitat restoration, which benefit both the economy and the environment.

- Non-Governmental Organizations (NGOs): NGOs play an essential role in advocating for environmental protection, community rights, and social equity within river basin management. These organizations bring specialized knowledge of local ecosystems, community needs, and sustainable practices, making them valuable partners in promoting sustainable water use. NGOs often act as intermediaries between communities and government agencies, helping to amplify the voices of marginalized groups. They also conduct research, raise awareness, and engage in policy advocacy, holding governments and businesses accountable for their environmental and social impacts. The contributions of NGOs are especially important in contexts where public oversight may be limited, as they provide transparency and push for sustainable water management policies.

- Local Communities: Local communities, including Indigenous groups, farmers, and fishers, are among the most directly affected stakeholders in river basin management. They rely on river basin resources for drinking water, food production, and cultural practices, and they possess deep knowledge of the local environment. Engaging local communities in water management is essential for ensuring that decisions are culturally respectful, environmentally sound, and economically viable. Community-based knowledge can provide valuable insights into seasonal patterns, ecosystem changes, and resource availability, which are critical for developing effective management strategies. By involving local communities, river basin managers can create policies that are more likely to be implemented successfully on the ground.

The diversity of stakeholders involved in river basin management highlights the need for inclusive and collaborative approaches. Engaging governments, businesses, NGOs, and local communities ensures that management strategies reflect a comprehensive understanding of social, economic, and environmental dimensions, leading to more balanced and sustainable outcomes.

Methods for Effective Stakeholder Engagement

Achieving meaningful stakeholder engagement requires well-designed participatory approaches, co-management frameworks, and efforts to build trust and foster collaboration. These methods help integrate stakeholder perspectives into the decision-making process, promoting inclusivity, accountability, and shared ownership of river basin management outcomes.

Participatory Approaches and Co-Management Frameworks

Participatory approaches involve actively involving stakeholders in the decision-making process, ensuring that their input is considered and that management strategies address their concerns. Participatory methods can range from public consultations and workshops to advisory committees and collaborative planning sessions. By creating forums for stakeholder interaction, participatory approaches enable river basin managers to build consensus, address potential conflicts, and generate community support for management initiatives.

Co-management frameworks are a more formalized approach to participatory engagement, establishing shared decision-making authority between governments, businesses, NGOs, and communities. In co-management arrangements, stakeholders work together in all aspects of river basin management, from setting goals and developing strategies to monitoring and evaluating outcomes. This approach fosters a sense of joint responsibility and accountability, as each stakeholder has a stake in the success of the management plan. Co-management is particularly effective in contexts where stakeholders have competing interests, as it encourages collaborative problem-solving and compromise.

The Mekong River Basin offers an example of successful co-management, where governments, NGOs, and local communities work together to manage fisheries, protect water quality, and address the impacts of hydropower development. By involving stakeholders in monitoring and decision-making, co-management promotes transparency, ensures that policies are grounded in local knowledge, and fosters a collaborative environment for addressing shared challenges.

Building Trust and Collaboration Among Stakeholders

Building trust is a cornerstone of effective stakeholder engagement, as it fosters open communication, reduces conflicts, and promotes long-term cooperation. Trust-building requires transparency, consistent communication, and respect for differing perspectives and values. One way to build trust is to establish clear communication channels, such as regular meetings, newsletters, and online platforms, where stakeholders can share information, provide updates, and raise concerns.

Collaboration among stakeholders can also be strengthened through joint projects, such as conservation initiatives, research studies, and capacity-building programs. These collaborative projects allow stakeholders to work together toward common goals, build relationships, and develop a shared understanding of river basin management issues. Conflict resolution mechanisms, such as mediation and negotiation, can help stakeholders address disagreements constructively, preventing disputes from undermining collaborative efforts.

In the Nile River Basin, the Nile Basin Initiative provides a platform for member states to collaborate on water management issues, build trust, and address shared challenges. Through joint projects, dialogue forums, and capacity-building programs, the initiative has helped foster a collaborative approach to water management, enabling stakeholders to work toward sustainable outcomes in a region marked by competing water demands.

Participatory approaches, co-management frameworks, and trust-building strategies are essential for fostering effective stakeholder engagement. These methods ensure that stakeholder contributions are meaningful, relevant, and integrated into decision-making processes, creating a sense of shared responsibility and commitment to sustainable river basin management.

Challenges in Stakeholder Engagement

While stakeholder engagement is critical to successful river basin management, it also presents several challenges. Power imbalances, conflicts of interest, and capacity gaps can hinder effective engagement and limit stakeholders' ability to participate meaningfully.

Power Imbalances

Power imbalances between different stakeholder groups often affect their ability to influence decision-making. For example, government agencies and large corporations may have greater financial resources, technical expertise, and access to decision-makers, allowing them to exert more influence over management outcomes. In contrast, local communities and NGOs may lack the resources or political connections needed to advocate effectively for their interests. These imbalances can lead to decisions that favor powerful stakeholders, marginalizing vulnerable groups and creating inequitable outcomes.

Addressing power imbalances requires creating platforms where all stakeholders can participate on an equal footing. This can include providing financial support, training, and capacity-building programs to empower marginalized groups, enabling them to engage effectively in the decision-making process. Transparent decision-making processes and accountability mechanisms are also essential for ensuring that all stakeholders have a voice and that management decisions reflect the interests of the broader community.

Conflicts of Interest

Conflicts of interest arise when stakeholders have competing priorities, such as economic development, environmental conservation, and social equity. Businesses may prioritize water use for industrial purposes, while local communities may prioritize water access for drinking and agriculture. These conflicting interests can lead to disagreements over water allocation, pollution control, and land use, complicating the decision-making process and creating tensions among stakeholders.

To manage conflicts of interest, river basin managers can facilitate dialogue and negotiation among stakeholders, helping them understand each other's perspectives and work toward mutually beneficial solutions. Conflict resolution mechanisms, such as mediation and consensus-building, provide structured approaches for addressing disagreements and finding common ground. Benefit-sharing arrangements, which involve distributing the economic and social benefits of river basin resources equitably, can also help align the interests of different stakeholders and promote cooperation.

Capacity Gaps

Capacity gaps, including limited technical knowledge, financial resources, and organizational skills, can hinder the ability of some stakeholders to participate effectively in river basin management. Local communities, for example, may lack the technical expertise needed to engage in water management discussions or advocate for sustainable practices. NGOs may face funding constraints that limit their ability to conduct research, educate the public, or monitor river basin activities.

Capacity-building programs are essential for addressing these gaps, as they provide stakeholders with the knowledge, skills, and resources needed to engage effectively. Training workshops, technical assistance, and financial support can empower stakeholders to participate meaningfully, enhancing the overall effectiveness of river basin management. International organizations, government agencies, and universities often play key roles in providing capacity-building support, enabling stakeholders to contribute their unique perspectives and expertise.

Challenges in stakeholder engagement, including power imbalances, conflicts of interest, and capacity gaps, highlight the need for inclusive, equitable, and supportive engagement strategies. By addressing these challenges, river basin managers can create a collaborative environment that supports sustainable and effective management.

Conclusion

Stakeholder engagement is essential for achieving sustainable and inclusive river basin management. By involving diverse groups in decision-making, river basin managers can ensure that management strategies reflect the needs, values, and expertise of all those affected by river basin policies. Inclusive engagement fosters collaboration, builds trust, and creates a sense of shared responsibility, contributing to the long-term sustainability of water resources.

Effective stakeholder engagement requires participatory approaches, co-management frameworks, and trust-building strategies that foster collaboration among governments, businesses, NGOs, and local communities. These methods ensure that stakeholder input is meaningful, relevant, and integrated into decision-making processes, leading to balanced and equitable outcomes.

Despite its importance, stakeholder engagement presents challenges, including power imbalances, conflicts of interest, and capacity gaps. By addressing these obstacles through capacity-building, transparent decision-making, and conflict resolution, river basin managers can create an environment where all stakeholders can participate effectively.

In conclusion, stakeholder engagement in river basin management is a dynamic and ongoing process that requires commitment, inclusivity, and adaptability. By fostering meaningful collaboration, river basin managers can develop strategies that support the sustainable management of water resources, protect ecosystems, and promote the well-being of communities for generations to come.

Chapter 10: Financing Sustainable River Basin Management

Sustainable river basin management is essential to support biodiversity, maintain ecosystem services, and ensure water availability for communities and industries alike. However, effective management requires significant financial resources, encompassing costs related to ecosystem restoration, infrastructure development, pollution control, monitoring, and governance. As pressures on water resources continue to grow, so too does the need for sustainable financing models that can support these activities over the long term. Identifying, securing, and managing financing is one of the core challenges for river basin management authorities, as funds must be allocated effectively to meet complex and evolving demands.

This chapter explores the financing needs of sustainable river basin management and examines various sources of financing, from traditional public funding to innovative mechanisms like payment for ecosystem services and green bonds. We also delve into case studies that illustrate successful financing models, showcasing examples of river basin projects that have achieved sustainable outcomes through well-structured funding. By understanding and applying diverse financing strategies, river basin managers can secure the resources necessary to support sustainable water use, protect ecosystems, and build resilience against climate change and other environmental pressures.

Funding Needs for Sustainable River Basin Management

Managing a river basin sustainably involves various activities, each with unique funding requirements. These include ecosystem restoration, infrastructure maintenance, pollution control, monitoring, and governance efforts. Each category has specific financial demands, requiring river basin managers to adopt a strategic approach to budgeting and resource allocation.

Costs Associated with Ecosystem Restoration

Ecosystem restoration is a critical component of sustainable river basin management, as it aims to rehabilitate damaged ecosystems, restore biodiversity, and enhance the natural functions of rivers and wetlands. Restoration efforts often include reforestation, wetland conservation, removal of invasive species, and riverbank stabilization. These activities require significant financial investment, covering costs associated with planning, labor, materials, and long-term monitoring to ensure the success of restoration projects. Additionally, ecosystem restoration often involves land acquisition, which can be expensive in densely populated or industrialized areas. Without adequate funding, restoration projects may fail to achieve their intended outcomes, underscoring the importance of secure, sustainable financing.

Infrastructure Development and Maintenance

Sustainable river basin management often requires substantial investment in infrastructure, including dams, reservoirs, irrigation systems, wastewater treatment facilities, and flood control structures. These assets play a crucial role in regulating water flow, controlling floods, and maintaining water quality. However, infrastructure development and maintenance are costly, as they involve capital investments, regular maintenance, and operational costs. Infrastructure projects also demand periodic upgrades to adapt to changing conditions, such as increased water demand, climate variability, and population growth. Without reliable financing, essential infrastructure may deteriorate, leading to inefficiencies, reduced water availability, and increased vulnerability to environmental risks.

Governance and Institutional Capacity

Effective governance and institutional capacity are foundational to sustainable river basin management. Governance activities include establishing regulatory frameworks, conducting research, monitoring water quality, and engaging with stakeholders. Funding is essential for hiring skilled personnel, conducting training programs, and investing in information technology systems that facilitate data collection and analysis. Governance efforts also involve coordination between different administrative levels and sectors, requiring resources for meetings, workshops, and stakeholder engagement initiatives. When governance institutions are underfunded, they may struggle to implement and enforce regulations, compromising the sustainability of river basin management efforts.

Monitoring and Data Collection

Continuous monitoring of water quality, quantity, and ecosystem health is essential for effective river basin management. Monitoring provides valuable data that inform decision-making, allowing managers to assess the impact of management strategies, identify pollution sources, and track changes in water availability. Monitoring requires specialized equipment, such as water quality sensors, flow meters, and remote sensing technology, as well as skilled personnel to analyze and interpret data. Funding is necessary not only for initial investments in monitoring technology but also for ongoing maintenance and upgrades. Without adequate financing, monitoring programs may become sporadic or incomplete, limiting managers' ability to make informed decisions.

The funding needs of sustainable river basin management are diverse and substantial, covering ecosystem restoration, infrastructure, governance, and monitoring. By identifying and addressing these needs, river basin managers can develop comprehensive financing plans that support the long-term sustainability of water resources and ecosystems.

Sources of Financing

Securing adequate financing for sustainable river basin management requires exploring a wide range of funding sources. These sources include traditional public and private financing, international development funds, and innovative mechanisms that incentivize sustainable practices. A diverse financing portfolio can help river basin managers secure the resources needed to meet their objectives and adapt to changing conditions.

Public and Private Financing

Public financing, often provided by national and local governments, is a primary source of funding for river basin management. Governments allocate funds through budgets, grants, and subsidies, supporting activities such as infrastructure development, pollution control, and ecosystem restoration. Public financing is particularly important for projects that provide broad public benefits, such as flood protection, water quality improvement, and habitat conservation. However, government budgets are often limited and subject to competing priorities, making it essential to supplement public financing with other sources.

Private financing, provided by businesses, investors, and philanthropic organizations, is increasingly important in river basin management. Companies with vested interests in water resources, such as agricultural producers and industrial users, often contribute to river basin management through corporate social responsibility (CSR) initiatives and public-private partnerships. For example, companies may fund projects that reduce water pollution, enhance water efficiency, or support local conservation efforts. Private financing can also take the form of investment in green bonds, which are designed to fund environmentally beneficial projects. By leveraging private financing, river basin managers can access additional resources and foster collaboration with businesses.

International Development Funds

International development organizations, such as the World Bank, Global Environment Facility (GEF), and United Nations Development Programme (UNDP), provide financial support for river basin management projects worldwide. These organizations offer grants, low-interest loans, and technical assistance to countries working to improve water management, protect biodiversity, and build climate resilience. Development funds often focus on projects in developing countries, where financing for river basin management is limited. For example, the GEF has supported projects that promote sustainable land management, restore degraded ecosystems, and enhance water governance in river basins across Africa, Asia, and Latin America.

International development funds play a crucial role in financing large-scale projects that require substantial investment and technical expertise. In addition to financial support, development organizations provide capacity-building programs, research, and technology transfer, enabling river basin managers to improve their skills and adopt best practices. By partnering with international organizations, river basin managers can access funding, technical assistance, and knowledge-sharing networks that support sustainable management.

Green Bonds and Sustainable Investment Funds

Green bonds are financial instruments designed to raise funds for environmentally beneficial projects, including those related to water management, ecosystem restoration, and climate resilience. Green bonds are issued by governments, corporations, and development banks, offering investors an opportunity to support sustainable initiatives while earning a return on their investment. Green bond funds are typically earmarked for projects that promote environmental sustainability, making them an ideal financing source for river basin management activities.

Sustainable investment funds, such as environmental, social, and governance (ESG) funds, also support river basin management by directing capital toward projects that meet specific sustainability criteria. These funds appeal to investors seeking to align their portfolios with environmental values, providing an additional financing source for water management projects. By leveraging green bonds and sustainable investment funds, river basin managers can access private capital while promoting the environmental and social benefits of their projects.

Innovative Financing Mechanisms

Innovative financing mechanisms, such as payment for ecosystem services (PES) and water pricing, offer creative solutions for funding river basin management. PES programs involve compensating landowners, farmers, or communities for adopting practices that protect water resources, such as reforestation, wetland restoration, or pollution reduction. PES programs create financial incentives for sustainable practices, helping to reduce pressures on river basins and support long-term conservation goals.

Water pricing is another innovative financing mechanism that involves charging users for water consumption based on the volume used. By setting prices that reflect the true cost of water provision, including infrastructure, maintenance, and environmental impacts, water pricing encourages efficient use and generates revenue for river basin management. Progressive pricing structures, where large users pay higher rates, can also promote equity and ensure that water resources are used responsibly. Together, PES and water pricing offer flexible and adaptable approaches to financing sustainable river basin management, addressing both conservation and economic goals.

By combining public and private financing, international development funds, green bonds, and innovative mechanisms, river basin managers can develop a diverse and resilient financing portfolio that supports sustainable water management.

Case Studies of Successful Financing Models

Examining case studies of well-funded river basin projects provides valuable insights into effective financing strategies. These examples illustrate how different funding sources and financing models have supported sustainable outcomes, offering lessons for river basin managers worldwide.

Murray-Darling Basin, Australia

The Murray-Darling Basin, one of Australia's most important agricultural regions, faces significant water management challenges due to competing demands, water scarcity, and environmental degradation. To address these issues, the Australian government implemented the Murray-Darling Basin Plan, a comprehensive management strategy that includes funding for water infrastructure, ecosystem restoration, and community engagement. Financing for the plan comes from a combination of government allocations, water trading, and environmental water buybacks. The government has also invested in water efficiency projects for agriculture, reducing water consumption and improving resilience to drought. The Murray-Darling Basin Plan demonstrates how public financing, water trading, and targeted investments in water efficiency can support sustainable river basin management.

Upper Tana-Nairobi Water Fund, Kenya

The Upper Tana-Nairobi Water Fund in Kenya provides a successful example of a public-private partnership and PES model that supports water conservation in the Tana River Basin. The fund was established by The Nature Conservancy in partnership with local water users, businesses, and government agencies. It compensates farmers and landowners in the Tana River Basin for adopting conservation practices, such as terracing, reforestation, and erosion control, which protect water quality and reduce sedimentation in downstream reservoirs. Funding is provided by private companies, including beverage and water utilities, who benefit from improved water quality and reduced treatment costs. This case study illustrates how PES and private sector engagement can create a sustainable financing model that aligns economic incentives with environmental goals.

Rhine River Basin, Europe

The Rhine River Basin, which flows through multiple European countries, provides an example of successful transboundary financing for river management. The International Commission for the Protection of the Rhine (ICPR) coordinates water quality, pollution control, and flood management efforts across the basin. Funding for ICPR initiatives is provided by member countries, who allocate resources based on their shared interests and responsibilities. The Rhine Action Program, an ICPR initiative, has led to significant improvements in water quality by funding pollution control measures, wastewater treatment upgrades, and habitat restoration. The Rhine River Basin demonstrates how international cooperation and shared financing can support sustainable management in transboundary rivers.

Colombia's Magdalena River Basin

In Colombia, the Magdalena River Basin faces challenges related to deforestation, mining, and pollution. The government has established a fund to finance sustainable development projects within the basin, supported by public funds, international grants, and private investment. This fund supports projects that promote reforestation, reduce erosion, and improve water quality. In addition to securing financing, the fund collaborates with local communities to build capacity and support sustainable land use practices. The Magdalena River Basin case study highlights the role of dedicated funds and community involvement in achieving sustainable river basin management.

These case studies illustrate the diverse financing strategies that can support sustainable river basin management. By adopting flexible, multi-source financing models, river basin managers can secure the resources necessary to protect water resources, restore ecosystems, and promote sustainable development.

Conclusion

Financing is a cornerstone of sustainable river basin management, enabling managers to undertake the projects and activities needed to protect water resources, support biodiversity, and ensure water availability. From ecosystem restoration to infrastructure development and governance, each aspect of river basin management has unique funding requirements that demand careful planning and resource allocation. By recognizing these needs and exploring diverse sources of financing, river basin managers can build a resilient and adaptable funding portfolio that supports long-term sustainability.

Sources of financing for river basin management include public and private funds, international development assistance, green bonds, and innovative mechanisms like PES and water pricing. These financing models offer flexible solutions that cater to the needs of different river basins, helping managers address environmental challenges while promoting economic development and social equity. Case studies of successful financing models, such as the Murray-Darling Basin Plan, Upper Tana-Nairobi Water Fund, Rhine River Basin, and Magdalena River Basin, provide valuable lessons for river basin managers worldwide, demonstrating the importance of diverse funding sources, collaboration, and strategic investments.

In conclusion, sustainable river basin management requires a commitment to securing and managing financing effectively. By leveraging multiple sources of funding, adopting innovative financing mechanisms, and learning from successful case studies, river basin managers can build a foundation for sustainable water management, ecosystem conservation, and climate resilience.

Conclusion: The Future of Sustainable River Basin Management

As we face increasingly complex environmental challenges, sustainable river basin management (SRBM) emerges as a crucial approach to secure water resources, maintain ecosystem health, and support communities. This book has highlighted the multifaceted strategies and solutions needed to achieve SRBM and illustrated how effective planning, technology, and governance can come together to foster sustainable management. In this conclusion, we will summarize the key takeaways, explore future directions, and provide a call to action for all stakeholders involved in shaping a resilient and water-secure future.

Key Takeaways from the Book

Throughout the book, a comprehensive framework for SRBM has been presented, encompassing strategic planning, technical innovations, regulatory frameworks, and cross-sectoral collaboration. The major strategies discussed include:

1. Integrated Water Resource Management (IWRM): This approach emphasizes coordinating the management of water, land, and related resources to maximize economic and social welfare while ensuring environmental sustainability. IWRM allows diverse stakeholders to work within a common framework, addressing the interconnectedness of water uses and impacts across a river basin.

2. Stakeholder Engagement and Collaborative Governance: Successful SRBM requires active participation from a wide range of stakeholders, including government agencies, local communities, businesses, and non-governmental organizations (NGOs). Collaborative governance ensures that management decisions reflect diverse perspectives, build public trust, and foster community buy-in for sustainable practices.

3. Nature-Based Solutions (NBS): Leveraging natural processes, such as wetland restoration, riparian buffer zones, and forest conservation, enhances ecosystem resilience and mitigates risks associated with flooding, drought, and erosion. NBS provide sustainable and cost-effective means to achieve SRBM by harmonizing human activities with natural systems.

4. Data-Driven Decision-Making and Monitoring: Advancements in digital tools, such as geographic information systems (GIS), remote sensing, and machine learning, have revolutionized SRBM. These technologies allow real-time monitoring of water quality, flow rates, and land-use changes, enabling informed, timely responses to emerging issues and helping stakeholders assess the impacts of management strategies.

5. Adaptive Management: River basins are dynamic ecosystems subject to climate variability and human pressures. Adaptive management provides a framework to continuously adjust policies, practices, and technologies based on observed outcomes, ensuring that SRBM can respond flexibly to changing conditions and new scientific insights.

6. Legal and Regulatory Measures: Strong policy frameworks are essential to enforce sustainable practices, protect water rights, and regulate pollution. Effective legislation balances the needs of development with environmental protection, creating enforceable standards for water use, waste discharge, and habitat conservation.

These strategies, when applied collectively, create a robust framework for SRBM, balancing the needs of humans and nature and laying the foundation for resilient and sustainable river basins.

Future Directions

To fully realize the potential of SRBM, ongoing research and innovation are essential. Climate change, urbanization, and population growth continually alter river basin ecosystems, requiring new knowledge and approaches to address emerging challenges. Research priorities should focus on:

- Understanding Climate Impacts: The hydrological cycle is sensitive to climate changes, and understanding how precipitation patterns, evapotranspiration rates, and river flows may shift is essential for future SRBM strategies.

- Developing Resilient Infrastructure: As extreme weather events become more frequent, it is crucial to invest in infrastructure designed to withstand such conditions. This includes building flexible reservoirs, flood control systems, and green infrastructure that can adapt to climatic uncertainty.

- Innovation in Water Treatment and Recycling Technologies: New technologies that allow efficient water recycling and reduce pollution at the source are vital to protecting water resources. Investments in treatment systems and pollution control can enhance water availability and improve ecosystem health within river basins.

Emphasizing the Importance of Resilience and Sustainability in Managing River Basins

A key future direction is integrating resilience and sustainability as core principles of SRBM. River basins that are resilient can recover from disturbances such as droughts, floods, and pollution events, while sustainability ensures that resource use does not deplete or damage ecosystems. Building resilience involves:

- Protecting and Restoring Ecosystems: Healthy ecosystems act as buffers, protecting communities from floods and filtering pollutants. Restoring degraded habitats and preserving natural landscapes within river basins enhances their ability to absorb shocks and adapt to changes.

- Promoting Sustainable Land Use and Agriculture: Agricultural practices are one of the largest consumers of water and contributors to pollution. Transitioning to sustainable farming techniques, such as precision agriculture, agroforestry, and organic farming, reduces pressures on river basins and supports biodiversity.

- Encouraging Sustainable Urban Planning: Urban areas contribute significantly to water stress through increased demand and pollution. Sustainable city planning can alleviate these pressures by incorporating green infrastructure, reducing impervious surfaces, and promoting water-efficient practices.

As we look to the future, ensuring that resilience and sustainability are prioritized in river basin management is essential for protecting ecosystems, supporting communities, and mitigating the impacts of climate change. Emphasizing resilience and sustainability also aligns with the broader goals of global frameworks, such as the Sustainable Development Goals (SDGs), particularly SDG 6 on clean water and sanitation, and SDG 13 on climate action.

Call to Action

The future of SRBM relies not only on strategies and technology but also on the collective commitment of society. Policymakers, practitioners, and the public each have a role in supporting and implementing sustainable practices for river basin management:

- For Policymakers: Governments at all levels need to enact policies that support SRBM, from protecting water sources to incentivizing sustainable land-use practices. Integrating SRBM into national and regional development plans, investing in ecosystem restoration, and promoting public-private partnerships can strengthen SRBM efforts and foster long-term resilience.

- For Practitioners and Water Managers: Those who manage water resources on the ground, including engineers, scientists, and local officials, should champion SRBM principles in their work.

Embracing adaptive management practices, advocating for nature-based solutions, and incorporating the latest technologies can improve the efficacy of SRBM initiatives and protect vital water resources.

- For the Public and Communities: Public awareness and engagement are critical to SRBM. Community-driven projects, citizen science initiatives, and environmental education can empower people to become stewards of their local water resources. Educating citizens about the importance of sustainable water use, pollution prevention, and habitat conservation creates a foundation for long-term, community-based resilience.

In conclusion, sustainable river basin management is not a one-time endeavor but a continuous process that requires adaptation, commitment, and a forward-looking perspective. As this book has shown, SRBM provides a comprehensive approach to addressing the complex challenges facing river basins today. By taking action now—whether through policy, practice, or public engagement—we can work toward a future where river basins are managed sustainably, ecosystems are preserved, and communities are safeguarded against water scarcity and environmental degradation. Sustainable river basin management offers a pathway to a water-secure future, one where people and nature can thrive together in harmony.

www.ingramcontent.com/pod-product-compliance
Lightning Source LLC
Chambersburg PA
CBHW071055240526
45469CB00006BD/2306